D1356055
Withdrawn From Stock
Dublin City Public Libraries

# The Science of Happiness

Gill Books
Hume Avenue
Park West
Dublin 12
www.gillbooks.ie

Gill Books is an imprint of M.H. Gill and Co.

© Brendan Kelly 2021
9780 7171 9005 8

Edited by Djinn von Noorden
Proofread by Jane Rogers
Indexed by Cliff Murphy
Printed by CPI Group (UK) Ltd, Croydon, CRO 4YY
This book is typeset in Minion Pro, 12 on 18pt.

This book is intended as general guidance only and does not in any way represent medical advice for individual persons. Readers are advised to contact their own general practitioners or other healthcare professionals for advice and guidance appropriate to their particular needs. This book does not in any way replace the advice and guidance that your own doctor or other healthcare professional can give you. If you are concerned about any of the issues raised in this book, be sure to consult your general practitioner. While every effort has been made to ensure the accuracy of the information and material contained in this book, it is still possible that errors or omissions may occur in the content. The author and publishers assume no responsibility for, and give no guarantees or warranties concerning, the accuracy, completeness or up-to-date nature of the information provided in this book.

*The paper used in this book comes from the wood pulp of managed forests. For every tree felled, at least one tree is planted, thereby renewing natural resources.*

All rights reserved.
No part of this publication may be copied, reproduced or transmitted in any form or by any means, without written permission of the publishers.
A CIP catalogue record for this book is available from the British Library.

5 4 3 2 1

# The Science of Happiness

The six principles of a happy life and the seven strategies for achieving it

BRENDAN KELLY

Gill Books

**Brendan Kelly** is Professor of Psychiatry at Trinity College Dublin, Consultant Psychiatrist at Tallaght University Hospital and Visiting Full Professor at the School of Medicine in University College Dublin. In addition to his medical degree (MB BCh BAO), he holds master's degrees in epidemiology (MSc), healthcare management (MA) and Buddhist studies (MA); and doctorates in medicine (MD), history (PhD), governance (DGov) and law (PhD). He has authored and co-authored over 250 medical and scientific papers, more than 500 other articles and 14 books, including *The Doctor Who Sat For a Year* (Gill, 2019). He contributes frequently to print and broadcast media. Born in Galway, Brendan now lives in Dublin with his wife, two children and a cat called Trixie. He is, for the most part, happy.

## Acknowledgements

I am very grateful to everyone who spoke with me and assisted with the preparation of this book. I particularly appreciate the support and guidance of my agent Vanessa O'Loughlin of the Inkwell Group (www.inkwellwriters.ie) and the superb team at Gill Books.

As ever, I remain deeply grateful to Regina, Eoin and Isabel, without whom none of this would be possible. I greatly appreciate the support of my parents (Mary and Desmond), sisters (Sinéad and Niamh) and nieces (Aoife and Aisling), as well as the ongoing advice and guidance of Dr Larkin Feeney, Dr John Bruzzi and Hector in the bakery. Also Trixie the cat, Terry the tortoise and all the fish deserve a mention. Hi!

In terms of happiness, I owe a special debt of gratitude to our friends in Italy who taught us that mushroom pizza in Rome and panzerotti in Bisceglie come very, very close to epitomising happiness for all. Grazie!

I am very grateful to Professor Veronica O'Keane, Ms Alison Collie and all of my amazing colleagues at Trinity College Dublin and Tallaght University Hospital.

I still benefit from the wisdom and advice of my teachers at Scoil Chaitríona, Renmore, Galway; St Joseph's Patrician College, Nun's Island, Galway; and the School of Medicine at NUI Galway.

Finally, a big thank you to my patients and their families, who have taught me so much about seeking well-being and maintaining good mental health.

This book is dedicated to Regina, for obvious reasons.

# Contents

INTRODUCTION

# What is Happiness?

Am I happy? This sounds like a simple question but, for some reason, we make it complicated. We strive to be happy, but we struggle to know whether or not we are truly happy. If we admit to being happy, we feel guilty. What if it does not last? Sometimes, there is comfort in sadness. Is that a kind of happiness too? The avoidance of disappointment? Is sadness actually happiness in disguise?

For most of us, the truth about happiness is relatively simple, even if we tie ourselves in knots trying to see it. We are either fairly happy and would like to be happier, or we are sad and would like to be happier. Either way, virtually all of us seek greater happiness in our lives even if we struggle to admit this clearly to ourselves.

And yet, we habitually act in ways that will not make us happier. We make the same bad choices again and again. We do not prioritise activities that increase our well-being. We feel guilty about pleasure. Why?

From a psychological perspective, there are three main reasons for our confused, counterproductive approach to happiness.

The first reason is that we dread disappointment even more than we desire happiness. We worry more about overpromising than underperforming. The desire to avoid disappointing other people and ourselves routinely trumps our unspoken desire to be happier. We dread failure far more than we covet success. This seriously limits our happiness.

The second reason why we hesitate to reach for happiness is that many of us are embarrassed about the desire to be happier. We feel that enduring a life of hardship and suffering somehow reflects well on us, compared to a life of enjoyment and ease. We are slow to discard our comforting myths of martyrdom. We rarely admit that we would like to be happier, and so we hesitate to take the steps that we know would boost our well-being. This, too, is deeply regrettable and greatly limits our prospects for future happiness.

The third reason why we fail to do simple things that would make us happier is that we struggle to balance the impact of our life choices with the impact of external factors that are outside our control. We swing between two extreme positions on this. When things are going well, we act as if we have 100 per cent control over our happiness. This is unwise because, when things stop going well, the responsibility overwhelms us. We are left in a state of paralysis, unable to make the simple changes in our day-to-day lives that we know would make us happier. We feel we have failed completely.

At this point, we swing dramatically in the opposite direction, deciding that factors outside our control are 100 per cent responsible for our sadness: where we were born, our family circumstances, our social situation and random life events. We think: 'I would be perfectly happy if I had been born in Finland, the happiest country in the world, but I wasn't, so there is nothing I can do.' This feeling of powerlessness when things are going badly is just as unhelpful as our exaggerated notions of our own power are when things are going well. We struggle to maintain a reasonable balance between these two extreme positions, acknowledging the extent of our own control on the one hand, and acknowledging the role of external circumstances on the other.

As a result of these three factors, our approach to happiness is conflicted, confused and counterproductive. This is a pity. Logically, most of us know about the many simple things we can do to make ourselves happier. A clear understanding of these steps, along with greater awareness of our tendency to avoid happiness, can help us to become happier, more fulfilled people. That is what this book is about.

'Happiness science' is a new field of research that has yielded fascinating insights about happiness and well-being over the past six decades. I find this area of study intriguing not only because happiness is interesting in itself, but also because happiness and well-being seem like impossible topics to study. How can researchers measure happiness? Surely, what I mean by 'happiness' might be very different to what you mean by 'happiness'? And my happiness changes constantly throughout the day. So how can scientists, psychologists, doctors, economists and philosophers perform reliable research into happiness?

It can be done. Most of the recent research in this field uses a simple, effective approach to measuring happiness. Every few years, the European Social Survey asks thousands of people in different countries a direct question: 'Taking all things together, how happy would you say you are?'[1] The person rates their happiness on a scale from 0 ('extremely unhappy') to 10 ('extremely happy'). This approach is refreshingly straightforward and, as a result, an overwhelming majority of people answer this question without hesitation. It just works.

One of the disadvantages of this simple way of assessing happiness at a given moment in time is that happiness can change quite quickly. If I rate my happiness at 7 out of 10 in the morning, I might rate it at 5 out of 10 in the afternoon. Researchers overcome this problem by asking thousands of people this question each time they do the survey, so that moment-to-moment fluctuations even out. In other words, the sheer volume of people asked the question negates the effect of factors such as time of day, the weather, random life events and so forth.

Another apparent disadvantage is that this happiness question is not 'objective'. Assessing happiness in this way is not like doing a blood test or a brain scan that someone else can check to make sure you got it right. The answer to the happiness question depends entirely on the person's *own* rating of their happiness at the moment when they are asked the question, rather than a more scientific assessment.

This is not a problem with the question itself, however, but a reflection of the nature of happiness. Happiness *is* subjective. A person's rating of their own happiness is the *only* rating that matters. No one can tell me how happy I am. Only I know that. So the best strategy is to ask me, simply and directly, how happy I feel I am – just like the researchers do.

That is exactly what I do every day in my clinical work. I am a psychiatrist, a medical doctor who specialises in the treatment of mental illness and psychological distress. General practitioners refer people to see me for many reasons: depression, anxiety, hyperactivity, hearing voices and a range of other symptoms. But, regardless of why someone comes into my consulting room, one of the first questions I ask is: how happy are you? I want to know to what extent a person's problems and symptoms disrupt their life, affect their mood or get them down. The best way to do this is with a simple, direct and open question. This question makes immediate sense to most people and they are always ready to answer it. This question just works.

There is a remarkable consistency in the results of happiness studies that ask this question. As we will explore, the factors associated with happiness turn out to be surprisingly consistent across many countries, even in cultures that, on the surface, appear very different from each other. Clearly, the simple, direct happiness question used in so many research studies maps onto a shared human understanding of what it means to be happy, no matter where you live in the world. We are more alike than we think.

This consistency of happiness research findings around the world brings me to the main reason why I am writing this particular book at this specific moment in my life.

In January 2020, shortly before my 47th birthday, economist David Blanchflower published a superb research paper looking at happiness across the human lifespan.[2] Who is happier, he asked, younger or older people? The pattern that Blanchflower identified in the data sets he analysed was remarkably constant across hundreds of different countries: human well-being reaches its lowest point at the age of 47 years. This same finding is reported in multiple studies around the globe with a consistency that is incredibly rare in any field of research. There is no doubt about it: 47 is the age of minimum well-being and maximum unhappiness on planet earth.

For obvious reasons, this research finding caught my eye at once. Was it true? Was this the low point for me? Was I now walking in the valley of darkness? Would everything get better from this point onwards? And, if so, why?

To be honest, I have always been interested in the idea of happiness. As a child, I presumed that everything adults did was aimed at increasing happiness – their own happiness and that of their families and friends. While this is true to a certain extent, I soon realised that much human activity is at least one step removed from happiness. We go to work not necessarily because work makes us happy, but to earn money for food, housing, holidays and other things that make us happy. With this in mind, is it reasonable to expect work itself to make us happy too? We tend to complain about work a lot, but we also expect it to give us satisfaction. Is that satisfaction different from happiness or is it the same thing?

And what about relationships? There is a common assumption that relationships are good for well-being, but where does that leave people who choose to be solitary? Are there different kinds of happiness? Or is happiness a single destination that can be reached by many paths?

Above all, I have always wondered what we can do in our own day-to-day lives in order to be happier. Is there a single set of guidelines, core principles or essential strategies that I can recommend to people who come to see me and apply in my own life so as to increase happiness, well-being and satisfaction?

This book is an exploration of this and various other questions posed by the happiness science of the past few decades. What follows is my effort to make sense of this growing field of research and turn its insights into practical advice for living happier lives.

This book takes a three-pronged approach to this task. The first two chapters summarise key findings from systematic research into happiness over recent decades in order to inform the rest of the book with as much science as possible. The third chapter brings the findings from the first two chapters into the areas of well-being, spirituality and psychology, and presents the six overarching principles of a happy life. The remaining chapters outline practical strategies to increase happiness in our day-to-day lives, focused on such areas as sleeping and waking, dreaming, diet, movement, activity, connection and finding healthy ways to lose ourselves and our worries in the world. As much as we want to focus on happiness, we sometimes need to become absorbed in other activities and simply let go. In the words of philosopher Henry David Thoreau: 'Happiness is like a butterfly; the more you chase it, the more it will elude you, but if you turn your attention to other things, it will come and sit softly on your shoulders.'

PART ONE

# The Science of Happiness

ONE

# Who We Are

Writing about happiness feels a little like dancing about architecture. We can certainly do it, but why should we? Would it not be better to spend our time being happy rather than researching it? Doing things that will increase our well-being instead of writing books and papers? Eating ice cream rather than trying to figure out why ice cream makes us happy?

The answer to this question is – infuriatingly – Yes and No. Yes, in the sense that we should always make space for the simple things that we enjoy: spending time with family and friends, going for walks, talking to the cat and eating ice cream. These things matter. They increase our well-being. We need more of them.

But the answer is also No because, deep down, we all know that while we enjoy these things, they are not enough to make us *happy*.



Leabharlanna Poiblí Chathair Baile Átha Cliath
Dublin City Public Libraries

No matter how much time we spend with people or cats, no matter how many walks we go on, and no matter how much ice cream we eat, we still need other things to make and keep us happy. For this reason, we need to both eat the ice cream and try to figure out why it makes us happy, how we can increase or share this happiness, and what else we can do to improve our well-being.

The happiness science of recent decades has focused on a wide range of factors that influence happiness. Some of these things, like how we spend our time, are largely under our control. Others, like what kind of childhood we had, appear much less controllable. What the research tries to do, however, is to examine precisely which of these factors truly shape our happiness, how they do so, to what extent, and what we can do about them. The results from the studies to date are sometimes predictable, frequently surprising and never less than fascinating.

This chapter starts our exploration of happiness science, both ancient and recent, by looking at research into some of the factors linked with happiness and related to *who we are:* our gender, age, genetic inheritance from our parents, upbringing and where we live. Many of these are factors over which we have limited control, but an awareness of how they impact on well-being can help us to navigate our worlds with a little more confidence, a little more insight and – hopefully – a little more happiness.

Let's start with gender.

## Are women happier than men?

It is an age-old question: are women happier than men, or are men happier than women? Recent decades have seen a great deal of research into happiness, depression and suicide in women and men in a determined attempt to shed light on this issue. The results are not always clear-cut, but they present interesting trends, some of which change over time.

To summarise the mental health research first, it is now clear that women are almost twice as likely as men to be diagnosed with depression, but men have a higher rate of suicide. While these findings are well established in relation to clinical depression and suicide, it is not clear what they mean for happiness in the population as a whole. The trends seem to point in different directions, with depression more common in women and suicide more common in men. In the end, these figures offer little guidance about happiness across the general population outside of those who are clinically depressed or experiencing a suicidal crisis. I see many people with these difficulties. They require much care and support if they are to overcome these issues and regain a sense of happiness and fulfilment in their lives, but it certainly can be done.

Looking at the population more broadly, a number of large-scale social surveys have now studied happiness across various populations and provide quite detailed information about the relationship between gender and happiness in the general population.

In the first place, it is important to remember that any statements about 'women' or 'men' will inevitably be gross generalisations. Both women and men are capable of happiness and unhappiness. Which of these states predominates will depend far more on the person's individual circumstances than their gender. In addition, there is growing recognition that the division between 'women' and 'men' is not as crisp as it was once imagined to be, as recent years have seen increased recognition of gender diversity and fluidity of gender identity. These are welcome developments: seeing the world in terms of 'women' and 'men' was always an oversimplification that obscured important issues just as often as it illuminated them.

Despite these caveats, the relationship between gender and happiness remains the subject of endless fascination in both the popular media and scientific research. As humans, we have an insatiable and seemingly unquenchable desire to describe all kinds of human behaviours in terms of gender, continually comparing women with men and men with women as if gender was ever the only factor that influences behaviour. Clearly, it is not: human behaviour is complex and multifactorial. Gender, if it is relevant at all, is only one factor among many and it is rarely the most important one.

Even so, gender remains one of the most common ways that we categorise or label each other in order to try to understand what we do, what we think and how we feel. With this in mind, let us look at some of the recent research about happiness and gender to see if it makes any sense. For the most part, it does.

When systematic studies of happiness began, in the middle of the twentieth century, results consistently showed that women rated themselves as happier than men rated themselves. This finding persisted up until the mid-1970s, when the wage gap between women and men began to narrow, educational opportunities increased for women, and various societal changes appeared to increase women's social freedom and economic possibilities. At that time, there was every reason to believe that these developments, although imperfect and still incomplete, would increase happiness and well-being for women, at least compared to men. But did they?

The best data to answer this question comes from the General Social Survey, which is a nationally representative survey of approximately 1,500 people (between 1972 and 1993) and 3,000 to 4,500 people (between 1994 and 2006) across the United States.[1] This survey, which is still performed today, contains detailed questions about subjective well-being and happiness. Looking at the data collected between 1972 and 2006, it appears that women's happiness *fell* substantially during this period, despite there being little change in men's reported happiness over the same time. Moreover, women's happiness fell both in absolute terms and relative to that of men, despite the apparent social and economic progress made by women over these decades.

This is a paradox. By most objective measures, the lives of women in the United States improved considerably over these 35 years, and other surveys indicated that women themselves regarded their lives as better too. But the happiness data from the General Social Survey show that happiness, in fact, shifted away from women and towards men over this period. This shift was evident across all industrialised countries, not just the United States.

These findings create an unexpected paradox. In terms of happiness, it seems that the chief beneficiaries of increased opportunities for women were men. Why?

Commenting on this research in the *Guardian*, Anna Petherick noted that women, despite living longer than men and gaining more political, economic and social freedoms, were not becoming any happier:

> So why is this? Evidence supports the idea that women's rights and roles in the home in the US and Europe have not moved in step with changes in the workplace. Therefore, because women with jobs often do most of the chores and childcare, they shoulder a dual burden that cuts into their

> sleep and fun. Long commutes are thought to make British women more miserable than British men because of the greater pressure on women to meet responsibilities at home as well as work.[2]

Social expectations matter too, as this 'dual burden' causes working women 'in Sweden, for example, to feel more miserable than their counterparts in Greece', owing to greater expectations about gender equality in Sweden.

In summary, then, women's increased work outside the home has occurred *in addition* to work inside the home, rather than *instead* of it. Social expectations of women arguably make the situation even worse in certain countries, with the result that happiness among women in developed countries has decreased rather than increased over past decades, especially compared to men.

These explanations, although dispiriting, appear to be true. This is commonly known as 'double jobbing', as women both work outside the home and carry most of the responsibility within the home too: cooking, cleaning, minding children, supporting relatives and looking after other household tasks. It appears that men now spend less time working and more time relaxing, and women spend more time working outside the home.[3]

These trends are also apparent in Europe. In 2010 the research group I led at University College Dublin used data from the European Social Survey to study happiness in over 30,000 people across 17 European countries.[4] Happiness was assessed using this question: 'Taking all things together, how happy would you say you are?' Each person rated their happiness on a scale from 0 ('extremely unhappy') to 10 ('extremely happy'). We found no relationship between happiness and gender. Mean happiness for both men and women across Europe was 7 out of 10. Our more detailed analysis found that greater happiness was associated with younger age, satisfaction with household income, being employed, high community trust and religious belief. But, just like in the United States, women in Europe no longer rated themselves as happier than men.

In a further analysis of European Social Survey data from 2018 performed for this book, I found that mean happiness is still equal among almost 17,000 men and 19,000 women across Europe, at an average of 7.4 out of 10.[5] The increase in happiness from 7.0 in 2010 to 7.4 in 2018 is attributable to the economic recovery after the global Great Recession that took place between 2007 and 2009. The fact that women's happiness remains at the same level as men's, however, bears continued testament to the relative decline in women's happiness compared to men's since the 70s.

So what is the solution? Is there any way to arrest the relative decline in happiness among women and ensure greater happiness for all?

There is now strong evidence that achieving greater gender equality is the key to both addressing the decline in happiness among women in developed countries and attaining greater happiness for men.[6] Advancing gender equality has been a key value in many societies for a long time, but progress is too slow. This is a pity. Not only is gender equality a vital goal in its own right, but happiness science adds yet another reason to pursue it with renewed vigour: gender equality will make *everyone* happier, women and men alike.

For the purpose of our exploration of happiness science here, the take-home message about gender and happiness is that women and men are now just as happy – or unhappy – as each other. But happiness is always multifactorial and these trends change over time. Just as women used to rate themselves as happier than men, men might soon rate themselves as happier than women. And, of course, there are many factors other than gender that are linked with happiness. One of these is age.

## Why is 47 the age of greatest unhappiness?

Who are happier, the young or the old? In many cultures, childhood is regarded as a time of innocence and happiness. In other cultures, later life is seen as a period of wisdom and well-being. Who is right? Does the pattern of happiness across the life-course differ between people, across countries or over time? Is there a pattern? And, if there is, what does it mean?

Some people struggle deeply with ageing. They avoid revealing how old they are. They wish they were younger. Birthdays upset them. They become less happy as they grow older. My age has never bothered me in the slightest, but maybe that will change in the future. Will I become happier or less happy as I age? To what extent do I decide this myself and to what extent do other factors decide it for me?

In the first place, it is important to emphasise that everyone's experience of happiness and unhappiness over the course of their lifetime is different. Some people have deeply unhappy childhoods owing to poverty, neglect, abuse and various adverse circumstances. Other people describe blissful childhoods but experience difficult periods in later life owing to life events, illness, loss, misfortune and random bad luck. Each person's trajectory is different, so any research about happiness and ageing will inevitably be general, focused on the 'average' human experience (which does not really exist) and challenging to apply to our own lives at any given point in time.

Despite these issues, and somewhat against the odds, it turns out that certain patterns are evident across many people's lives and it is interesting to think about these in relation to our own lives. For example, a great number of people find that their priorities change over the course of their lifetimes, often shifting from short-term happiness in the moment when one is younger to quieter, sustained pleasures that increase well-being in later life. We tend to mellow as the years fly past, becoming less interested in hectic late-night clubbing and more taken with gardening, walking and – in my case – sitting quietly, talking to the cat.

Two main conceptions of well-being are relevant here. The first is 'hedonic' well-being, which refers to seeking out experiences that maximise pleasure in the moment; e.g. taking an impulsive trip to Las Vegas rather than saving up for a new garden shed. The second concept is 'eudaemonic' well-being, which refers to a sense of fulfilment in life or a feeling of quiet satisfaction in progressive achievements over time; e.g. patiently contributing to a pension fund rather than following your favourite band on tour across Europe for the summer.

Both approaches to well-being are necessary in order to achieve a reasonable balance between living in the moment and feeling secure about the future. But the balance we achieve between these two ends of the well-being spectrum varies greatly both between people and within the same person over time. Everyone's trajectory is different. Not everyone abandons nightclubs in later life and many young people enjoy gardening.

So what does all of this mean for well-being and happiness over the course of a lifespan? Does a growing emphasis on life satisfaction rather than heedless hedonism mean that well-being and happiness diminish over time? Or does the opposite occur, as greater age brings enhanced focus on the present moment rather than planning for a future that you now know rarely turns out how you expect it to? The relationship between happiness and age has long been an area ripe for definitive research.

In 2020 economist David Blanchflower drew together some hundreds of data sets about well-being and happiness from around the world and came to some clear conclusions about the relationship between well-being and age,[7] confirming the finding that happiness and well-being are U-shaped across the human lifespan. In essence, most of us start out in life with relatively high levels of happiness and well-being in childhood and adolescence. Regrettably, these decline as we enter our 20s and 30s and reach their lowest points in our late 40s. Happily, though, the curve then bends upwards again as we enter our 50s. Our levels of well-being and happiness are restored and continue to curve upwards as we move through later life.

To summarise: we start life happy; we become sad; and then we become happier again.

There are two remarkable aspects to Blanchflower's findings. The first is how consistent the results are across so many countries, and the second is how strong the findings are: the midlife dip in happiness is substantial. This level of consistency and reliability is rare in this field and is worth thinking about for a moment.

First, consistency. Blanchflower drew upon systematic comparisons across no fewer than 109 data files and found evidence of the U-shape in some 132 countries, including 95 developing countries, even after taking account of other factors that affect happiness such as education, marital status and employment. Averaging across the

257 country estimates from developing countries, he found that 48.2 years is the age of minimum well-being in these settings. Averaging across the 187 country estimates he obtained for advanced countries, Blanchflower found that 47.2 years is the age of minimum well-being in these countries.

The consistency of Blanchflower's findings is truly astonishing, especially across so many nations with contrasting social structures, cultural traditions and economic circumstances. No matter where we go in the developed world, it seems, our late 40s will likely be the low point for well-being and happiness in our lives. The only significant exception might be countries with life expectancies that are shorter than the world average, for which the limited evidence available suggests that the U-shape still pertains, but the dip might occur at a different age.

Second, the pattern that Blanchflower identified is not a minor one. The midlife dip in well-being and happiness is not just consistently seen around the world, it is also big in magnitude. This is not a minor dip in mood: it is a steep curve downwards that is three decades in the making, is notably low when it happens, and takes another decade or two to come back up. It is not a minor blip. There is no doubt about it – 47 is the age of minimum well-being and maximum unhappiness, and the magnitude of this dip cannot be ignored.

To summarise: the midlife dip in well-being and happiness occurs everywhere around the world and is deep enough to affect our lives in significant and noticeable ways. This immediately prompts two more questions: Why does this occur? And is there anything we can do about it?

There are many reasons why the middle period of life is the least happy. In the first instance, this is often the period of maximum responsibility. Many people are either working outside or inside the home (or both) or have the stresses of not working and looking for a job. For those who work outside the home, the mid-40s are a time when they are either promoted or they are not. Career-wise, this is a make-or-break phase in a person's working life that brings long hours in the office, intense workplace politics, frustrated ambition and either disappointment or additional responsibilities. Either way, well-being is adversely affected.

Outside of work, many people are paying a mortgage, supporting and looking after a family, contemplating the care needs of ageing relatives and subject to infinite informal obligations to family and friends. More recently, social media platforms have amplified this intensity, creating an additional hamster wheel of compulsive communication that simply adds to the pressure. Finally, midlife is a period when many people develop problems with alcohol or other drugs that cause additional complexities of their own, diminish their usual coping mechanisms and magnify the effects of other challenges in their lives. Midlife is a tough time. It's no wonder we're so unhappy.

Can we do anything to alleviate this malaise? Happily, we can. Jonathan Rauch, author of *The Happiness Curve: Why Life Gets Better After Midlife*, recommends normalising our midlife dissatisfaction by recognising that this dip is an acknowledged phenomenon, other people feel this way and other people get through it.[8] He recommends interrupting our self-criticism, developing greater focus on the present moment and reaching out for support when needed. He suggests that we see this as a step-by-step process, rather than trying to solve the midlife dip in one big leap. We are incremental creatures. Sustainable change takes time. This is wise advice, especially during our difficult late 40s.

But perhaps the best advice of all is simply to sit with the knowledge that this period of maximum unhappiness will come to a natural end. The happiness curve turns upwards again as we enter our 50s. Patience is key. It can be difficult to cultivate patience when life is so busy, full of worries and concerns, and seemingly out of control. Simple practices help greatly. If your life feels like a runaway train, try to set aside time for a quiet, absorbing activity that lets the rest of the world gently melt from your mind: reading, running, listening to music. We all need more of these activities and we will return to these kinds of steps later when we focus on practical strategies to increase happiness in our day-to-day lives.

For now, the key message from the happiness science is that well-being and happiness are significantly related to age: we start out happy, our well-being declines as we enter midlife and we reach the low point of well-being and happiness around the age of 47 years. From that point onwards, things improve: well-being and happiness rise again as we enter later life. Of course, random life events can and will disrupt this trajectory in all kinds of ways, but this pattern tends to hold true for most people in most countries for most of the time. All told, it is a remarkably consistent finding.

This realisation offers valuable consolation if you, like me, are in your late 40s and living through the phase of your life when happiness tends to be at its lowest. This, too, shall pass.

## Do we inherit a genetic set point for happiness?

We all know happy people. These are people who seem to float through life with a smile on their faces, no matter what. They keep things in proportion and deal with life's problems calmly as they arise. They are, at times, annoying, but they also seem happy. How come?

Of course we all know, deep down, that these people face just as many challenges and difficulties as anybody else, but – somehow – they don't let their problems crush their spirit. It appears, for all intents and purposes, that these people were born happy and life does little or nothing to dent their optimism. Does this really happen? Are some people just born happy? Or do these people learn how to be happy along the way? Is it a mix of both?

This is something that has puzzled philosophers, theologians and writers for many centuries. The essential question is this: how much of our happiness is in our nature, our biology and the way that we are built, and how much is attributable to nurture, the environment and our personal life choices?

The answer to this question has become much clearer over the past three decades. Research from the mid-1990s shows that, despite life's ups and downs, many people's capacity for happiness remains relatively constant.[9] Four inner traits are significantly associated with happy people: high self-esteem, a feeling of personal control, optimism and extraversion.

The net effect is that while people have a clear ability to adapt to both bad news and good news, we also have a set point around which our happiness fluctuates over time, but from which it never departs too far or for too long. We spend our lives hovering around this point. This is an extraordinary finding that immediately prompts another question: how is such a set point determined and can we influence it during our lives?

The answer is enormously surprising and it centres on the relationship between happiness and genes. Genes are the basic physical and functional units of heredity.[10] Genes are the biological code we inherit from our parents that gives us some, but not all, of our parents' characteristics. Humans have between 20,000 and 25,000 genes. Most genes are identical in all people, but a small proportion – less than 1 per cent – are slightly different between people. This 1 per cent of our genes is what makes each of us into the unique individual that we are. Might this 1 per cent include some genes for happiness?

Once it was clear that people have a built-in propensity to hover around a certain level of happiness for their entire lives, the search began for a 'happiness gene'. Of course, genes are complex entities and only one element among multiple factors that make us who we are. The idea that a specific gene is the sole determinant of anything as complex as happiness is, for the most part, misleading. Nevertheless, genes are usually a key part of the mix and better understandings of genetics have contributed significantly to human science and psychology over recent decades.

The genetic closeness of twins has contributed to important studies of the role of genes. In relation to happiness, there has been interesting research on 2,310 members of the Minnesota Twin Registry, a birth-record-based registry of twins born in Minnesota from 1936 to 1955.[11] Researchers studied these twins as they went to school and became adults, and so they know how far the twins progressed in school, their approximate family income, their marital status and their socio-economic status, based on their occupations. It is an astonishing study.

The Minnesota sample includes both monozygotic twins (who share precisely the same genes, i.e. 'identical' twins) and dizygotic twins (who do not share precisely the same genes but are as genetically close as regular siblings). There is also research about twins who were reared together and twins who were reared apart. This mix of twin pairs allows researchers to make subtle estimates about how much of a given trait, such as happiness, appears to be attributable to genes and how much is attributable to other factors in the environment, such as education, income, marital status and employment.

Based on analysis of the Minnesota twin study, it appears that neither socio-economic status, educational attainment, family income, marital status nor religious commitment account for more than approximately 3 per cent of the variance in well-being between individuals in the study. Between 44 per cent and 52 per cent of the variance in well-being is associated with genetic variation. As a result, while transitory fluctuations in well-being depend on passing life events and random circumstances, the midpoint of such variations is largely genetically determined. These results strongly support the idea that we have a happiness set point that is largely determined by our genes.

This rather startling conclusion has been borne out in more recent research, which confirms the substantial role of genes in happiness and well-being. The next step is to try to identify any specific genes that are involved, figuring out how they work and seeing if their function can be safely boosted in order to increase happiness. These tasks are not simple, but some progress is being made.[12]

While it is still not entirely clear how any specific genes that might be identified could contribute to well-being on a biological level, findings strongly support the overall role of genetics in shaping happiness. Recent research uses intensely sophisticated genetic technology that was unknown as recently as ten years ago. The pace of change in this field strongly suggests that there will be further insights into the relationship between genes and happiness over the coming years.

So if up to 50 per cent of the variance in our happiness is genetically determined, what does this mean for us as we try to be happier?[13] Does it mean that we should try harder, since half of our happiness is apparently under our control? Or does it mean that we should devote less effort to being happy, because half of our happiness is already genetically determined? You could look at this either way.

In the end, the message that you take from this research likely depends on your personal temperament. Ironically, temperament is another human characteristic that is substantially shaped by genetics. But regardless of your temperament, the first thing to remember is that having genes that predispose you to happiness will also predispose you to happy experiences, which will in turn generate greater happiness. As a result, the estimate that 50 per cent of happiness is genetically determined is likely to be an overestimate, although it is impossible to estimate just how much of an overestimate it is.

The best approach is to acknowledge that genes have a substantial impact on happiness, but they still account for less than half of the variance in happiness between individuals. This leaves over half of the variance in happiness either unexplained or attributable to the various other factors that we are discussing here: gender, age, health, life circumstances, life choices and so forth.

Psychologist Maureen Gaffney, in her book *Flourishing: How to achieve a deeper sense of well-being, meaning and purpose – even when facing adversity,* describes the genetic predisposition to happiness not so much as a set point but a likely range for happiness.[14] Our genetic inheritance certainly has a role to play, and may indicate certain parameters within which our happiness is likely to lie, but genes do not act directly to produce or prevent happiness. Humans are hugely complex biological systems and we have substantial conscious control over what we do, how we feel, the way we live our lives and, ultimately, how happy we end up being.

The eighteenth-century English poet William Cowper wrote that 'happiness depends, as nature shows, less on exterior things than most suppose'. Both exterior and interior factors shape our happiness. We used to underestimate the interior, genetic part of this equation, but it is possible that we overestimate it now, in light of recent research.

The facts are that genes certainly matter to happiness (it would be strange if they did not), but other factors matter too. Taken together, these other factors play a significantly greater role than genes in determining our happiness across our lifespans. Childhood experiences are one of these other factors that help to shape our well-being and we will consider these next.

## Does childhood matter to happiness?

Childhood is a time of extraordinary opportunity. It is a period of great happiness for some, a time of deep suffering for others, and a phase of significant instability for the many who are caught in situations of conflict, deprivation or abuse. It is now accepted as a matter of faith that childhood experiences have a significant influence on an adult's well-being and happiness. Indeed, many people seek to explain virtually every problem they have as an adult by reference to something that happened in their childhood. Are they correct to do this? Does childhood really matter so much? And, if it does, is there anything we can do about it?

The great thinkers of history have never hesitated to provide advice about parenting and children. Aristotle emphasised education: 'Those who educate children well are more to be honoured than they who produce them; for these only gave them life, those the art of living well.' Noting tensions between children and parents, Plato pointed out that it is the job of children to differ from their parents: 'Don't force your children into your ways, for they were created for a time different from your own.' This is not always easy for parents to accept. I see many adults who are entirely baffled by the fact that their children's worldview differs from their own. I point out, as gently as I can, that their parents likely went through a similar period of difficulty when they were younger.

Despite a plenitude of advice from all kinds of sources over many centuries, however, childhood remains a time of uncertainty, difficulty and distress for many. There is no doubt that our early life experiences – be they happy, sad, or a mixture of the two – echo strongly into our adult lives. As a result, recent decades have seen considerable research into the precise effects of childhood on adult mental health and well-being. The ultimate aim of such work is to identify ways to support children and families to ensure better outcomes and greater happiness for all.

To summarise, children need to be loved and to be firmly attached to an adult, often a parent.[15] The nature and tone of childhood matter deeply. If a child is subjected to physical abuse, the child is at increased risk of becoming an abusive parent themselves. If a child experiences emotional abuse, they are three times more likely to develop depression as adults. Parental addiction to alcohol or drugs can also contribute to mental health problems in the child. Consistent, positive parenting and parental love are essential if a child is to become a well-adjusted, happy adult.

Even in the absence of abuse or addiction, however, conflict in the family home is an especially corrosive force. People who grow up in conflict-ridden homes report substantially lower levels of psychological well-being as adults compared to other people.[16] Even the long-term effects of divorce depend on the level of conflict in the home. If there is a lot of conflict leading up to the divorce, children report higher levels of well-being as young adults if their parents subsequently divorce, compared to children whose parents stay together and continue to fight. Much research confirms this finding: conflict in the home is especially destructive of future happiness and is to be avoided at all costs. Put simply, a high level of conflict within a family home diminishes a child's prospects for future happiness for life.

As well as presenting an opportunity to diminish or limit future happiness, however, childhood also presents a unique opportunity to boost the future well-being of children. Research suggests that promoting good social and emotional functioning in childhood is especially important for a child's future happiness. There are, for example, clear and significant associations between social and emotional skills in kindergarten and positive outcomes in young adulthood, including better education, higher likelihood of stable employment, lower likelihood of criminal activity, less substance use and better mental health.[17]

Why does developing good social and emotional skills in childhood matter so much and impact so deeply on future well-being? The answer lies in the close relationship between social networks and well-being. In a sense, we all know this already: having a good circle of friends and acquaintances contributes to well-being, resilience and human thriving. Many people whom I see in my clinical work lack this support structure, for a variety of reasons. Rebuilding it is often a key step on the path to recovery.

This is clearly demonstrated in the Framingham Heart Study, a long-term follow-up study that started in Massachusetts in 1948 with 5,209 people and went on to enrol most of the children of that original group – and then their children too.[18] Analysis of 20 years of data shows how happiness was distributed across this evolving social network, with clusters of happy and unhappy people clearly visible. People who were surrounded by many happy people and those who were central in the network were more likely to become happy in the future. Analysis over time showed that this increase in happiness resulted from the spread of happiness through the network and not simply a tendency for people to associate with similar individuals. Happiness, it seems, is contagious.

As a result, it is clear that any person's happiness depends on the happiness of other people with whom they are connected. Happiness, like health, can be usefully seen as a collective phenomenon rather than an individual one. This means that a deeper understanding of social networks is vital if we are to better navigate the challenges facing our world.[19] One of these challenges is achieving greater happiness and well-being for all. Research to date, taken as a whole, shows that developing social and emotional skills in childhood is an essential step along the road towards achieving this goal.

So in the end, what does all this research teach us about the relevance of childhood to the future happiness and well-being of adults? The most important lesson is that the ancient philosophers were (of course) correct: childhood matters greatly to future happiness. Each child needs to be loved and to enjoy a childhood that is free from abuse, conflict and parental addiction to alcohol or drugs. Consistent, positive parenting is essential. The development of social and emotional skills is especially important so that children have fewer problems as they grow up. In addition, we now know that happiness is an intrinsically networked phenomenon that spreads from person to person, so gaining good interpersonal skills in childhood helps people form part of happy social networks as adults.

In order to achieve these goals, psychologist Martin Seligman argues that we should teach well-being in schools so as to promote better learning, increase life satisfaction and decrease anxiety and depression.[20] Seligman outlines various school-based initiatives along these lines, focused on teaching assertiveness, creative brainstorming, relaxation, decision-making and assorted other coping skills. Positive outcomes from these programmes include diminished anxiety, reduced symptoms of depression, fewer conduct problems and better health-related behaviours such as self-care and exercise.

The usefulness of these interventions is rooted in the clear relationship between childhood and happiness. Childhood matters deeply. To return to Greek philosophy, Aristotle is sometimes credited with the maxim: 'Give me a child until he is seven and I will show you the man.' The fact that this insight is also attributed to St Ignatius of Loyola, co-founder of the Society of Jesus (Jesuits), simply reflects its manifest truth: what children experience and learn from the earliest age has a substantial impact on their future lives, including their experience of well-being and happiness as adults.

Of course, other things matter too. As we have seen from the happiness research so far, childhood is by no means the sole determinant of adult well-being. As a result – and tempting though it is – it is incorrect to relate all of one's adult problems to issues stemming from childhood. While childhood is certainly important, myriad other factors are relevant too, including genetic inheritance (although how this works is still obscure), life choices and, intriguingly, where we live in the world. Literally, where on earth is the greatest happiness to be found?

## What is the happiest country in the world?

Is it possible to compare happiness across different countries around the world? Or is each country simply too different from the next to make any meaningful comparison? When someone in Peru says that they are happy, are they talking about the same thing as someone in Nigeria? When someone in Canada describes well-being, are they referring to the same concept as someone in New Zealand? Cambodia? Mexico? Are there any consistencies or national differences at all? Or are happiness and well-being just too personal to permit cross-border comparisons?

While I have always lived in Ireland, I have spent time in various other countries: the United States, India, China, Japan, Russia and many European and Scandinavian nations. While some features of human life are constant wherever you go, there are also remarkable differences in lifestyles, values and ideas between certain countries. Terminology differs, too, especially in relation to happiness. *Arbejdsglæde* means happiness at work in Denmark, while *dolce far niente* is the sweetness of doing nothing in Italy.[21] *Mai pen rai* means 'don't worry' in Thailand and *þetta reddast* is resilience in Iceland. Do all of these concepts map onto the same thing – happiness? Or are they fundamentally different?

The first point to note is that all of these conceptualisations of happiness exist in all of these countries, but different countries place varying emphases on each one and describe them slightly differently. There is, despite the many terms used, a clear commonality in the experience of happiness across national borders, even if each country chooses a different way to express certain aspects of that happiness. In many senses, we are more alike than we think, despite the linguistic and cultural differences that make both life and travel so interesting.

The second point worth noting is that virtually all of the happiness science from around the world presented in this book is based on surveys that use a question similar to that in the European Social Survey: 'Taking all things together, how happy would you say you are?'[22] Each person rates their happiness on a scale from 0 ('extremely unhappy') to 10 ('extremely happy'). This question performs well in diverse countries around the globe, providing further evidence that there is a shared concept of happiness that is relatively constant, despite regional or linguistic differences in how this experience is most commonly expressed in various languages.

On this basis there is now considerable research interest in comparing happiness across national borders, finding out which country rates itself as the happiest, and then figuring out why this is the case. Maybe other countries have lessons to learn from the happiest ones?

With this in mind, the Sustainable Development Solutions Network publishes the annual *World Happiness Report*, authored by a group of leading experts in happiness and well-being. The report contains a wealth of data and information about happiness and well-being around the world, but most attention is usually paid to the rankings of happiness that are presented as a league table of countries. Naturally, everyone wants to know which country is the happiest in the world and where their own country stands on the list.

How are the rankings calculated? The *World Happiness Report* uses data from the Gallup World Poll of people aged 15 years and older.[23] This research is performed all over the world with nationally representative samples within each country where possible. The survey represents a staggering 95 per cent of the world's adult population. The countries' happiness rankings are based on answers to the main life-evaluation question in the poll. This is known as the 'Cantril ladder' and it is quite similar to the methodology of the European Social Survey. The key question in the Gallup World Poll asks respondents to think of a ladder, with the worst possible life being a score of 0 and the best possible life being a score of 10. Respondents are asked to rate their current lives on this 0-to-10 scale.

So what do the results show? In 2020 the *World Happiness Report* rankings of happiness included 153 countries.[24] The top five happiest countries, starting with the happiest, were Finland (with an average score of 7.81 out of 10), Denmark (7.65), Switzerland (7.56), Iceland (7.50) and Norway (7.49); the United Kingdom was thirteenth (7.16), Ireland was sixteenth (7.09), the United States was eighteenth (6.94), China was ninety-fourth (5.12) and India was one hundred and forty-fourth (3.57). The five unhappiest countries, starting with the unhappiest, were Afghanistan (2.57), South Sudan (2.82), Zimbabwe (3.30), Rwanda (3.31) and Central African Republic (3.48).

As ever, these rankings provide much food for thought. The most striking aspect is the role of wealth: rich countries tend to be happier. Happiness depends on a great many other factors as well as wealth, but severe poverty and political instability make it difficult for countries to achieve high happiness ratings. All of the unhappiest countries in the *World Happiness Report* are either poor or politically unstable or both. There is now strong evidence that happiness increases over time if economies grow and incomes rise. As a result, poverty and global inequality are undoubtedly the key ethical and political issues facing the world at present. Imbalances in happiness and well-being are just one manifestation of these problems.

Looking to the top end of the happiness table, the predominance of Scandinavian or Nordic countries is remarkable: four of the five happiest countries in the world in 2020 are Scandinavian or Nordic, and Sweden comes in seventh (7.35). In 2019 the top five happiest countries were very similar: Finland (7.77), Denmark (7.60), Norway (7.55), Iceland (7.49) and the Netherlands (7.49).[25] Again, Sweden came in seventh (7.34). The five unhappiest countries were also similar: South Sudan (2.85), Central African Republic (3.08), Afghanistan (3.20), Tanzania (3.23) and Rwanda (3.33).

Scandinavian or Nordic countries dominate right throughout the *World Happiness Report*. The global ranking of cities, for example, includes Helsinki (Finland) and Aarhus (Denmark) in first and second places. Copenhagen (Denmark), Bergen (Norway) and Oslo (Norway) are fifth, sixth and seventh, while Stockholm (Sweden) is ninth. Overall, Scandinavian or Nordic cities account for six of the top ten cities in the world, based on how positively inhabitants currently evaluate their lives. The other four, non-Nordic, cities in the top ten are Wellington (New Zealand), Zurich (Switzerland), Tel Aviv (Israel) and Brisbane (Australia), in third, fourth, eighth and tenth places, respectively. Washington (USA) comes eighteenth; Dublin (Ireland) is twenty-second; and London (United Kingdom) is thirty-sixth.

Why do Scandinavians consistently rate themselves as happier than virtually everyone else on the plant? What is it about Finland that makes it the world's happiest country? Why are the Danes; the Norwegians; the Icelanders so happy?

The short answer is that nobody knows, but the absence of any clear reason has not dampened a tsunami of speculation about possible explanations. Common suggestions include the relative wealth of many of these countries, high average incomes, relatively long life expectancy, strong government support services, good social connections and notable equality in certain aspects of life.

But many other countries have some or all of these factors too, and their citizens do not rate their happiness as highly as the Scandinavians do. Perhaps there are more intangible reasons that are not captured by standard measurements of wealth, stability and social progress?

Some commentators relate the Scandinavian happiness phenomenon more to Nordic culture and a certain way of living than to any specific, measurable indices. These more intangible reasons might include different approaches to community responsibility, a particular understanding of public space or an appreciation of social bonds that differs from that in many other countries: more permissive and less prescriptive, but still imbued with deep solidarity and a sense of responsibility.

The 2020 *World Happiness Report* provides some possible explanations for the dominance of Nordic nations in its rankings, in a valuable chapter titled 'The Nordic Exceptionalism: What Explains Why the Nordic Countries are Constantly Among the Happiest in the World.'[26] This chapter notes that since 2013 the five Nordic countries – Finland, Denmark, Norway, Sweden, and Iceland – have all been in the top ten, with Nordic countries commonly occupying the top three spots. Why?

There are several possible explanations for this, including the generosity of the welfare state in these countries, the high quality of government and institutions, low levels of income inequality, relative autonomy, freedom to make life choices, high levels of trust in other people, greater social cohesion and a general ethos of equality. Various historical and cultural factors might also be relevant, although these are more difficult to pin down.

Ironically, it appears easier to explain why certain countries are at the bottom of the happiness table (poverty, political instability, etc.) than to explain why certain other countries are at the top. This is true of happiness in individuals as well as countries; we always feel we can explain unhappiness more readily than happiness. But if the happiness science of recent decades proves anything, it is that happiness can indeed be studied and explained, at least in part – happiness science has much to teach us.

Before we all book our tickets to move to Finland, however, it is wise to pause for a moment and reflect on the fact that world happiness rankings – like much happiness research – come with caveats. There are huge individual variations in happiness, even in Nordic countries, and many factors other than location impact on a person's well-being. If someone with an unhappy childhood flees to Scandinavia in search of the good life, they bring echoes of their early life with them. They also bring their genetic inheritance and they still face the dip in happiness at 47, even if coffee at a sidewalk café in Helsinki might help to take the edge off it some days.

Happiness is complex, although not quite as mysterious or unpredictable as we like to imagine. Maybe all we need to do is to look – and think – a little harder.

TWO

# What We Do

Like many people, I often imagine that my choices determine how happy I am. While this is partly true, the previous chapter focused on what happiness science tells us about the relationship between happiness and *who we are* – our gender, age, genes, upbringing and place of residence. Apart from where we live, we have little control over these factors, although we can try to understand them better and appreciate their role in our lives. This can help with self-knowledge, acceptance and – I hope – happiness.

By way of contrast, this chapter focuses on *what we do* and the critical choices we make in our lives. To what extent do these shape our happiness? Some of our key decisions relate to starting a family (does having a baby make you happier?), income (how much money do you need?), employment (does work make you happy?), health (how important is physical well-being?) and values (do religion and politics matter?). Findings from studies on these topics help us to recognise both the importance and the limits of our life choices in shaping our happiness and well-being. Let's start with babies.

## Does having a baby make you happier?

Does having children make the parents happy? Marcus Tullius Cicero, a Roman statesman and philosopher, had no doubts: 'Of all nature's gifts to the human race, what is sweeter to a man than his children?' When Cicero's beloved daughter, Tullia, died unexpectedly in 45 BCE, he was devastated: 'I have lost the one thing that bound me to life,' he wrote. Any bereaved parent will understand how Cicero felt. There are few losses deeper than the death of a child.

Several centuries later, Thomas Paine, the eighteenth-century political activist and philosopher, prioritised his children's well-being over his own during a particularly troubled period in American history: 'I prefer peace. But if trouble must come, let it come in my time, so that my children can live in peace.' Again, many parents will understand Paine's desire to make sacrifices so that his children might live better, safer, happier lives. This is one of the deepest instincts of parenting across the natural world: to protect one's young.

While the love of parents for their children cannot be doubted, however, it is considerably less clear whether having children actually makes the parents any happier. In other words, while the rhetoric surrounding children is almost uniformly positive, focused on the joys of family life, recent decades have seen increased evidence that having children is by no means a reliable path to happiness. The truth is a lot more complicated and considerably more interesting than that.

So what does the research tell us?

First, there have been numerous studies of parenthood and marital satisfaction. Taken together, these studies show that parents describe significantly lower marital satisfaction than non-parents,[1] or couples with children are less satisfied with their marriages compared to couples who do not have children. This effect is more pronounced among women compared to men, and women's marital satisfaction is especially low in the year after giving birth. While 62 per cent of childless women express high levels of satisfaction with their marriages, this falls to 38 per cent of women with infants. The age of the child is not related to marital satisfaction among men.

It is possible that this decline in marital satisfaction might relate to the effects of the transition to parenthood, gender-specific reasons (in light of greater dissatisfaction among women than men), socio-economic factors (there is more dissatisfaction among people who are better off, strangely) or generational effects (because dissatisfaction appears to increase over time). To offset this, it is helpful if potential parents set reasonable expectations for parenthood, share responsibility more equally, and bear in mind that a drop in marital satisfaction might be more than offset by other gains, such as children adding new meaning to life.

The gender differences outlined in this research are evident across many studies in different countries, with significant evidence that children reduce marital happiness for women, but not for men, and having children living at home reduces marital happiness for both men and women between the ages of 50 and 70 years.[2]

These differences between women and men might relate to gender expectations that are apparent from the time of childbirth. One study from Ghana reported that 46 per cent of women feel less happy than their husbands following childbirth; 9 per cent feel equally happy and 45 per cent feel happier.[3] Intriguingly, the greater the difference in risk-taking between wife and husband, the happier the wife: men are happier if the couple have similar preferences in relation to trust and altruism, but women are happier if there are larger differences between them in risk-taking.

This is a fascinating, counter-intuitive finding because it is decidedly unclear why women should feel greater happiness if they do not share preferences with their husbands in relation to areas such as risk-taking, trust and altruism, all of which seem central to parenting. What is clear from this study, however, is that the happiness of women differs significantly from that of men following childbirth, whatever the underlying reasons might be.

Of course, gender is only one of many factors shaping happiness after having children. There are also differences over time, not least because happiness is never static for long, especially following childbirth. Broadly speaking, happiness increases during the year before birth and remains elevated in the year of the birth, but then quickly returns to pre-birth levels.[4] Comparing women with men, women have steeper rises in happiness before and after the birth, and then steeper falls a year later (presumably because their initial rise was steeper to begin with). In other words, having a baby brings happiness for the year before and the year after the birth, but happiness then returns to what it was a year before the birth.

People who have children at older ages or have more education experience a higher positive happiness response to a first birth, and, while having two children increases happiness, having a third does not. These are 'average' figures and many people may have different experiences, but these patterns appear to hold true for many people.

Are these patterns the same all around the world? A short-lived increase in happiness at the time of the birth, followed by diminished marital satisfaction, especially among women?

One survey of 22 countries in the Organization for Economic Cooperation and Development (OECD) found that non-parents are indeed happier than parents in some countries, but the reverse is true in others.[5] Countries in which non-parents are happier than parents include the United States, Ireland, Greece, the United Kingdom, New Zealand, Switzerland, Poland, Australia, Denmark, the Netherlands, Israel, the Czech Republic, Germany and Belgium. Among these countries, the biggest difference between non-parents and parents is in the United States, followed by Ireland.

In sharp contrast, countries in which parents are happier than non-parents include Portugal, Hungary, Spain, Norway, Sweden, Finland, France and Russia, with the greatest difference between parents and non-parents seen in Portugal, followed by Hungary. What accounts for these differences? The research suggests that subsidised child care and rights to paid leave from work might have the greatest impact on parental happiness, among the policy interventions studied. These more generous family policies are associated with smaller disparities in happiness between parents and non-parents around the world.

Finally, as parents get older, do children bring more or less happiness compared to the earlier years? One survey of 55,000 middle-aged and older adults across 16 European countries found that while marriage and children are associated with greater well-being and lack of depressive symptoms in this slightly older age group, the positive association with children is only evident after the children leave home.[6] In other words, when parents are middle-aged or older, resident children do not have a positive effect and any association between children and well-being only holds true if the children leave home.

Commenting on this theme in the *Observer*, Kenan Malik noted that 'children can make you happy', but 'only once they've left home':

> It's the latest in a pile of recent studies that have sought to measure parenting and happiness. While the results have been mixed, most suggest that parents are less happy than non-parents. The very question 'Do children make parents happy?' would have seemed odd a generation or two ago. Having children was simply what you did.
>
> Possessing greater reproductive choice has been a boon, especially for women. But the way we think of choice has also distorted our perception of happiness and of the significance of children.[7]

There is, in fact, good evidence that we have never truly understood either happiness or its relationship to having children.[8] We have always focused disproportionately on the positive aspects of parenthood rather than the negative ones. Perhaps that is why we continue to have children despite the many complexities of family life.

The lessons from the happiness science of recent decades could inject some much-needed rationality into our thinking about having children. This is important: I often see people who either did not think before deciding to have children or are disappointed that having children proved more complicated than they imagined.

In a nutshell, the research shows that parenthood boosts well-being for two years around the time of a birth (up to a maximum of two births); well-being is lower among parents compared to non-parents (in some but not all countries); and older parents whose children have left home experience greater well-being than those whose children still live with them. The evidence also suggests that the happiness deficit among parents in certain countries might be remedied by more equitable distribution of responsibilities among parents and policies that support subsidised child care and rights to paid leave from work.

Children will not automatically make us happy in a lasting way, but they have unique potential to boost our well-being, provided circumstances facilitate this.

So once you've thought it through, fire ahead.

## How much money do you need to earn?

Does money make us happy? Most of us imagine that additional income would be useful in our day-to-day lives or to fund specific projects. But many of us also have a sneaking suspicion that being wealthy would not necessarily make us happy. We have all heard stories about unhappy millionaires and everyone has read about the pressures that sudden, immense wealth can bring to new celebrities.

Even so, a great majority of people – including me – would be willing to give fabulous wealth a try, just to see how happy or unhappy it might make me. If it didn't work out, I would give the money back. I promise.

Luckily there is now a vast amount of research available to guide us on the role of money in relation to happiness. Some of the findings from this work are contradictory or inconclusive, and most researchers are at pains to emphasise the importance of factors other than income and wealth in shaping human happiness. Even so, in the midst of the clamour and confusion, certain trends are now relatively clear and provide useful guidance as we think about the role of money in our lives.

Research demonstrates, beyond all doubt, that it is difficult or impossible to be happy if you struggle to have enough water, food, shelter or medicine to live a safe and healthy life. Additional income is most valuable when it raises people out of poverty and enables them to meet their basic human needs.[9] This makes sense. It means that addressing global poverty is one of the key steps we can take to increase the happiness of the human race. There are, of course, many other reasons to eliminate poverty, such as improving health, prolonging life and achieving greater justice, but increasing happiness is yet another reason to focus on this goal. Extreme poverty makes happiness exceptionally difficult to achieve.

But once a person is raised out of poverty, how important is increasing income for increasing happiness? And, if there is a link, is there any limit to how happy income and wealth can make a person?

In 2010 one study analysed data from the Gallup Organization's polls relating to a total of 136,839 people, including just over 1,000 people from each of 132 countries.[10] These nations vary widely in terms of political structure, stability, economic development and culture, and therefore provide a good picture of the human race as a whole. This study found – predictably – that well-being is inversely related to unmet basic needs. If a person's fundamental human needs are not met, their well-being is likely to be low. In addition, greater income is associated with more positive feelings and fewer negative feelings; i.e. greater income is indeed associated with greater well-being. Interestingly, however, *societal* income is also linked with life evaluations, in addition to individual income.

So what does this mean? Does increasing income make us happier in the ways that matter? And, if so, is there is an optimal income that maximises the chances of happiness? Plenty of research addresses this issue, including one study of 450,000 responses to the Gallup-Healthways Well-Being Index, which is a daily survey of 1,000 residents of the United States conducted by the Gallup Organization.[11] This study showed that income and education are closely related to life evaluation, but health, care-giving, loneliness and smoking are relatively stronger predictors of daily emotions. In addition, while emotional well-being is indeed linked with income, there is no further benefit beyond an annual income of approximately $75,000. Low income is linked with both low life evaluation and low emotional well-being, but while high income might buy life satisfaction, it does not buy *happiness.*

Is $75,000 the magic number – the target income for maximum well-being with minimum hassle? Perhaps not. Other research paints a slightly more complex picture, suggesting that satiation occurs at $60,000 for positive emotions, $75,000 for negative emotions and $95,000 for life evaluation, albeit with some variation across world regions, as satiation occurs later in wealthier places.[12] There is no appreciable increase in subjective well-being beyond these incomes and, in certain parts of the world, incomes beyond the satiation points are associated with *lower* life evaluations.

The diminishing value of added income as you earn more makes sense. If a person who has no money suddenly earns $95,000 in a year, their well-being and happiness are likely to increase quite dramatically. But if someone who already earns $10 million in a year suddenly earns an additional $95,000, it is unlikely to have any effect on their positive emotions, negative emotions or life evaluation.

In fact, research suggests that additional income at this level might reduce well-being rather than increase it.

Where does all this research get us? First, the research so far confirms the common-sense idea that income matters to happiness. If our basic needs are not met, it is difficult to be happy. Second, increasing income correlates with greater well-being and happiness, but only to a certain point. Income beyond the equivalent of $95,000 per year is unlikely to bring significant additional happiness to most people. Third, it is perfectly possible that income beyond the $95,000 threshold will diminish rather than enhance our well-being, so perhaps we should be careful what we wish for.

All of these findings are based on broad population studies that conceal large amounts of individual variation. Some people are very happy despite being poor. Some millionaires are miserable. Some billionaires are ecstatic. Well-being and happiness depend on so many factors that it is misleading to place particular emphasis on any single component such as income or wealth.[13] The combination of factors that shape a person's happiness will differ between individuals, between countries and over time.

It is also true to say that financial well-being matters to everyone, at least to some degree. That is why research about income and happiness features routinely in newspapers, magazines and websites, and sparks endless discussions on television, radio and web-forums. In addition to pointing to the relationship between income and well-being, however, and establishing the optimum income for happiness, research on this theme has several other dimensions that deserve attention.

First, let's consider the importance of our income as it compares to the incomes of others. There is strong evidence that the well-being we derive from our income is linked not only to how much money we earn, but also how much money other people earn compared to us.[14] As a result, my life satisfaction increases if a pay rise improves the ranking of my income compared to the income of others. The problem with this is that if I increase my rank relative to others, their ranks must decrease. Thus, my additional life satisfaction from a pay rise appears to come at the expense of the life satisfaction of others.

This problem is not, however, insurmountable. There is evidence from the United States that progressive taxation – a higher tax rate for higher earners – helps to resolve this issue. One analysis of the General Social Survey found that income inequality was smaller in years when income tax was more progressive.[15] During these years Americans in the lowest 40 per cent of the income distribution tended to be happier and the richest 20 per cent were no less happy as a result. In this scenario, it is likely that income rankings stayed more or less the same, as the incomes of the poor went up and the rich lost out so little that it did not impact on their happiness at all.

Second, the source of a person's income, and how they spend it, is linked with the happiness they derive from it. For example, millionaires who earn their wealth are significantly happier than those who inherit it.[16] Of course, many millionaires derive their fortune from a combination of earnings and inheritance, but research suggests that the greater the earned proportion, the greater the happiness associated with the wealth. This appeals to an innate human desire to feel that we have earned our good fortune, rather than attributing it to random chance or a simple accident of birth.

Spending also matters. We like to help others. One group of researchers asked 632 Americans to rate their happiness, report their income and estimate how they spent their money in a typical month.[17] These researchers found that spending more of one's income on other people was associated with greater happiness both across populations and over time. They randomly assigned some study participants to spend a small amount of money ($5 or $20) either on themselves or on others, and participants rated their happiness before and after their spending. Participants who were assigned to spend the money on others experienced greater happiness than those assigned to spend the money on themselves. The value of spending, therefore, lies not only in what we get for our money, but also the amount of virtue that we perceive in our spending habits.

Third, there is a complicated, unresolved relationship between national income and national happiness (i.e. income and happiness at the level of countries rather than individuals). While happiness varies with income both within countries and between countries at a given moment in time, it appears that happiness does not necessarily increase as a country's overall income rises over time. This is known as the 'Easterlin paradox', outlined by Richard Easterlin, professor of economics at the University of Pennsylvania, in 1974.[18] Almost half a century later, this paradox is still widely debated. Perhaps the chief point to take from this debate is that research findings about individual happiness do not necessarily translate readily to the level of entire countries. In addition, there are differences between rich and poor countries in relation to at least some of these parameters, further complicating the study of the happiness of nations rather than the happiness of individuals.

Fourth, individual-level factors such as trust, fairness and personality are all linked with the relationship between income and happiness and need to be considered when drawing any conclusions from the research discussed here. Data from the United States, for example, shows that Americans trust each other more and perceive other people to be fairer in years when there is less income inequality.[19] In relation to personality, it appears that income has a particularly significant influence on life satisfaction among people who focus on outer-oriented concepts, such as satisfaction, enjoyment and fulfilment, rather than people who focus on inner-oriented concepts, such as stoicism, virtue and tranquillity.[20] Other concepts and emotions such as jealousy are relevant if income inequality increases beyond a certain point, leading to disillusionment, jealousy and unhappiness.[21]

Most of us understand much of this already from our own lives and those of our families and friends. I certainly see the importance of income every day in my clinical work as a psychiatrist, especially when I am at outpatient clinics in disadvantaged parts of Dublin. People who struggle to make ends meet often have great difficulty remaining happy. But while more money is not always the answer, it is often part of the solution, as research in this area shows. We should not place all our hopes in cold, hard cash, but, below a certain point, the absence of money makes many things much more difficult.

To summarise, recent research confirms that poverty makes happiness difficult to sustain and that earning more money increases happiness, but the benefits diminish sharply beyond an annual income of around $95,000. How we perceive our income compared to the incomes of other people matters a great deal, as do how we obtain and spend our money. Personality matters, too. In the end, money has significant potential to increase happiness, but there are limits to how much it can do for us, and other factors matter even more: one of these is work, which appears to hold substantial value above and beyond the income that it generates for us.

## Does work make you happy?

Unlike cats, humans dislike idleness. While cats appear content to sleep for long periods, wake slowly, stretch their muscles and then return to sleep, humans are usually happiest when doing things. This tendency is undoubtedly the root cause of a great deal of our unhappiness. Many of the world's misfortunes, if not all of them, can be attributed to humans' desire to do things when we might be better advised to do nothing – or simply sleep, like cats. And yet we act, even when we should refrain. How does this relate to happiness?

Benjamin Disraeli, who twice served as prime minister of the United Kingdom in the nineteenth century, argued that 'action may not always bring happiness, but there is no happiness without action'. Disraeli certainly personified a particular kind of action with his lengthy political career and prolific output of novels, poetry, drama and non-fiction. If action truly produces happiness, Disraeli should have been very happy indeed.

But he wasn't. Disraeli suffered from lethargy, disillusionment and depression for a prolonged period, despite his impressive output and extraordinary achievements. While there were undoubtedly many reasons for Disraeli's psychological malaise, his comment raises the interesting question of whether or not a particular kind of activity is required in order to produce happiness. Is it only meaningful work, rather than frenzied activity, that makes us happy? Or is the link other way round, and is it happiness that makes us work, rather than work that makes us happy? Aristotle said that 'pleasure in the job puts perfection in the work'. Does this mean that happiness fuels good work, rather than good work fuelling happiness?

Happily, the relationships between work, employment and well-being have been studied extensively over the past few decades and new research findings shed considerable light on many of these issues. Let us consider unemployment.

Unemployment is associated with lower life satisfaction and substantially diminished happiness. Most of us instinctually know this, because unemployment is clearly linked with loss of income, loss of role and, for many, loss of crucial social connections. Unemployment can also bring relationship problems, loss of meaning and crises of identity. All told, unemployment is singularly inconsistent with happiness, and numerous studies strongly support this view.

One of the most detailed studies in this field examined data from the British Household Panel Survey (a study covering a random sample of approximately 10,000 people in 5,500 British households every year), the German Socio-Economic Panel (a similar survey in Germany) and the European Community Household Panel (which covers all countries in the European Union).[22] The findings were clear: the effects of unemployment on life satisfaction are negative and significant in all three settings. In addition, people do not habituate to being unemployed over time and unemployment has deleterious effects on well-being in both the short term and the long term.

This being established, the next question is: what can be done to address this problem? The most obvious solution is re-employment, getting people back into jobs. But does that really undo the problems caused by unemployment?

One research group looked at this question by examining data from the German Socio-Economic Panel to see if people who were unemployed returned to their previous happiness set point after they regained employment.[23] They looked at data from 3,733 people and found that while unemployment produced the expected decrease in life satisfaction, and re-employment produced a shift back towards people's previous set points, most people did not return completely to the levels of satisfaction that they had enjoyed prior to losing their jobs, despite being re-employed.

This finding suggests that the negative effects of unemployment persist even after people regain employment. It also suggests that factors other than the mere fact of working are relevant to the life satisfaction associated with work. This makes sense. Having been unemployed in the past is likely to affect feelings of job security well into the future.

This is probably a useful point for me to pause for a quick reflection on hypocrisy, which is a real risk for anyone writing a book about happiness. As a medical doctor, I have not yet been unemployed. I have held a range of jobs, some of which were more fulfilling than others, but I cannot pretend to understand what it is like to be unemployed. I have seen many people who struggle to find and sustain work and who demonstrate the negative effects of unemployment on mental health. Even though I have probably seen thousands of people with these problems, I have not experienced them myself. This is, of course, inevitable: when writing a book about a topic as broad as happiness, it would be unrealistic and hypocritical to contend that I have personal experience of every aspect of the topic. I do not. What I can do, however, is distil relevant research findings from the scientific literature, recount the experiences of people I have met, and share any personal experiences that are relevant to the topic at hand.

I enjoy my work, albeit with the inevitable good days and bad days. The scientific literature about different job types, however, suggests that certain jobs are systematically likely to deliver different levels of satisfaction, depending, of course, on the person and the job involved. Blue-collar workers, in areas such as construction and manufacturing, report lower levels of happiness than managers, executives and professionals, in all countries around the world.[24] The position of the self-employed is more complex. In most developed nations, being self-employed is associated with both higher life evaluation and more negative emotions such as worry and stress: being self-employed is both rewarding and stressful at the same time, it seems.

But one thing is clear overall: unemployment is associated with substantially reduced well-being. This effect is so consistent across all studies and in all regions because the value of employment stems not only from the money earned, but also from the social status, interpersonal relations, daily structures and achievement of goals associated with paid work. Unemployment impacts on all these benefits, no matter where you live. Moreover, the drop in well-being associated with losing a job is greater than the gain associated with getting one.[25] In other words, losing a job has a bigger impact than finding work.

Other things matter too, of course, and there are strong relationships between well-being and housing, for example, including security of tenure in the private rented sector. Commenting on these themes in the *Guardian*, writer and philosopher Julian Baggini notes that most media responses to research findings on these issues generally ignore matters relating to job quality, incomes and the need for security in housing. 'Such misplaced attention is par for the course,' Baggini argues, and 'allows us to neatly depoliticise the issue' of happiness:

> Whichever way you look at it, there is no escaping the conclusion that increasing well-being across society requires joined-up, long-term policy efforts [...] well-being flourishes when people have freedom and security. A correct reading of the well-being data puts the ball very much in the government's court.[26]

Overall, the research firmly supports the view that a strong focus on employment is vital if governments are to create policies to boost population well-being. Unemployment is unequally distributed across society and is strongly associated with poor physical and mental health.[27]

But unemployment, it seems, has a negative impact not only on the well-being of those who are unemployed, but also on that of other people in society who are negatively affected by the 'spill-over' effects of high unemployment rates, *even if they are still employed themselves.*[28] This finding is supported by evidence from multiple countries and leaves no room for doubt: unemployment is uniquely corrosive to the well-being of the unemployed *and* those in employment.

Of course, jobs vary greatly and not all work makes a positive contribution to well-being. Highly engaged employees can better satisfy their basic psychological needs at work compared to employees with low engagement or high levels of burnout.[29] The nature and degree of motivation associated with jobs also determines the levels of satisfaction and happiness to be gleaned from work.[30] In addition, there can be two-way relationships between job satisfaction and life satisfaction, with each influencing the other in complex ways.[31]

Overall, however, the research is clear and unambiguous: unemployment creates huge unhappiness among the unemployed and an increase in the national unemployment rate makes *everyone* unhappy, employed and unemployed alike.

Disraeli was right: action matters greatly and employment matters most. Governments, please take note.

## How important is physical well-being?

Is there a link between physical health and happiness? If so, how strong is the association and what can we do to optimise the benefits that good physical health can bring to our sense of mental well-being?

Perhaps the first issue to address here is the extraordinary tendency to regard physical and mental health as somehow separate from each other. As a psychiatrist – a medical doctor who specialises in the treatment of mental illness and psychological distress – I am continually intrigued by the tendency to regard mental health as entirely separate from physical health. When pressed, many people agree that there is a link between the body and the mind, but most stop short of regarding physical and mental health as simply two sides of the same coin. Why?

Our heads are clearly attached to the rest of our bodies. As a result, everything that we eat, drink, do and experience with any part of our body will have an impact on our brain in one way or another. Each of us is a single, embodied organism. It is helpful if we maintain an awareness of this fact and avoid drifting into the false belief that our mental life is significantly separate from our physical one. It is not.

That said, what does recent research tell us about the relationship between physical and mental health, psychological well-being and happiness? The first point to note is that we commonly identify certain links between our health and happiness.[32] People can, of course, adapt to physical limitations to a certain extent, but many struggle to adapt to chronic pain and mental illness. In 2013 our research group at University College Dublin examined European Social Survey data about happiness in Ireland and found that satisfaction with health has a consistently strong association with happiness.[33] Indeed, happiness is often more closely associated with health than with income. Good health matters to mental well-being even more than income does, according to some studies.

This makes instinctual sense and there is much research to support this position. One analysis of the National Survey of American Life, for example, examined the association between happiness and self-rated physical health among African-American men and found that men who report being happy also report better physical health.[34] In this study happiness was also associated with being married, being employed and earning more than $30,000 per year. Happiness was not associated with any particular age or level of education in this sample.

Analysis of data in the World Values Survey from fifteen countries across five continents confirms that poor health is strongly associated with unhappiness and dissatisfaction right around the world.[35] This survey also found that happiness is associated with being married and younger and older ages, consistent with the U-shape of happiness across the lifespan (see Chapter 1). Satisfaction is also commonly associated with subjective relative income and being employed, which echoes our earlier conclusions about income and employment.

Interestingly, a growing body of research examines the idea that reduced happiness is not only a consequence of ill-health, but might also contribute to poor health in the future.[36] There are several potential mediating links including physical activity, dietary choices and a range of biological processes possibly linked to inflammation and hormones. Some of the research on these themes is inconsistent, but it is increasingly difficult to deny the evidence that links happiness today with physical well-being in the future, supporting the idea that happiness *predicts* future health, at least in part.

A seminal study published in the *Lancet* in 2015 strongly supports this view.[37] The paper reported findings from 9,050 older adults in the English Longitudinal Study of Ageing who were followed up for an average of 8.5 years to see if eudemonic well-being (finding meaning and purpose in life) is associated with better future health. The average age of participants in this study was 65 years. The researchers divided the group into four quarters based on well-being and found that 29 per cent of people in the quarter with the lowest level of well-being died during the follow-up period, compared to just 9 per cent of those in the quarter with the highest level of well-being. On this basis, it appears that a higher level of well-being is associated with longer life in people around the age of 65.

While these results do not prove that well-being *causes* better health and a longer life, they provide persuasive evidence that a high level of well-being is associated with good health in the future, and not just in the present or the past. This suggests that well-being might both improve health in the future and prolong life in older adults, although it does not definitively prove these points.

Other studies provide further support for these ideas. In Australia data from almost 10,000 people in the Household Income and Labour Dynamics Survey show that happiness and life satisfaction are positively associated with good, very good and excellent health three years later.[38] Happiness and life satisfaction are also linked with the absence of long-term limiting health conditions and higher physical health levels in future years. These findings suggest that happiness and life satisfaction at the very least *predict* good future health and might even contribute to causing it.

Despite a need for more information in certain areas, we can now say, with some certainty, that subjective well-being *can* influence health.[39] Possible links between subjective well-being and good physical health include cardiovascular mediators, the immune system, the endocrine system (hormones), genetic factors, wound healing and health behaviour. These are all possible, promising links in need of more research.

Of course, health itself also influences well-being and happiness, so this causes further complexity when studying the direction of these relationships. But it is now clear that subjective well-being *can* influence health and that further work is needed to figure out precisely when this occurs and whether interventions aimed at increasing subjective well-being will also improve physical health and longevity. Clearly, the use of positive psychology to improve future health appears to be a promising strategy.[40]

So there is a link between physical health and happiness: good physical health appears to contribute to happiness, and happiness appears to contribute to good physical health. Owing to these close associations, it is not surprising that certain activities and lifestyle choices contribute to *both* physical health *and* happiness, albeit possibly at different times and in slightly different ways. Let's look at diet, for example.

One 2016 study looked at the evolution of well-being and happiness after increases in the consumption of fruit and vegetables by examining the food diaries of 12,385 randomly selected Australian adults.[41] This study found that increased consumption of fruit and vegetables was associated with enhanced happiness, life satisfaction and well-being over the following two years. We already know that healthy foods are associated with better physical health, but those benefits can take decades to accrue, so it might be useful to show people evidence of the *happiness gains* from a better diet – gains that occur sooner than the physical health benefits, which can seem quite distant.

For me, foregoing ice cream and chocolate biscuits gives me a short-lived sense of having made a virtuous choice, but it is difficult to weigh this against the immediate energy burst that sugar brings. Now, if I take on board the evidence that a better diet makes me happier today as well as healthier in the future, this might change my decision-making considerably. Hopefully, I will make fewer visits to the biscuit tin and thus avoid the sugar highs and sugar crashes this brings. The more evidence I see about this, the greater the likelihood that I will change my habits.

Mental and physical well-being are not separate entities. Each can fuel the other, and both are influenced by similar factors, including a good diet and physical activity.[42] Our heads are not separate from our bodies. This means that, for better or worse, our happiness and mental well-being are tightly wedded to our physical health and physical well-being and each matters deeply to the other.

## Do religion and politics matter?

According to the Buddha, 'There is no path to happiness. Happiness is the path.' While comments such as this draw a clear link between happiness and a certain form of spirituality, it is not entirely clear how one finds the right path and, once found, how one sticks with it. Does Buddhism makes you happy? Does any religion?

Many people grow up in specific religious traditions and it can be difficult to separate feelings about religion from feelings about childhood in general. I was brought up in the Roman Catholic faith, of which I remain part, although I have developed an interest in Buddhism over the years. It is difficult to figure out to what extent this background affects my happiness. I'm just me. It is virtually impossible to dissect out the various different parts of me (or any person) in order to answer questions like this. Does the fact that I have an interest in religion (in all religions, if I'm honest) mean that I am more likely to be happy? Or less likely? Does following a religion have any relevance to happiness at all? Maybe it doesn't matter?

Virtually all religions and spiritual traditions speak about happiness and it is commonly imagined that religious belief and spiritual practices can help to make at least some people happier. This idea is remarkably persistent despite many religions focusing much of their energy on themes of guilt, punishment and exclusion, rather than happiness. So does religion make you happy?

Research on this topic has increased greatly over the past two decades. Between 1872 and 2000 over 1,200 research studies containing original data about the relationship between religion or spirituality and health were published.[43] Between 2000 and 2010, however, an additional 2,100 studies appeared, which hugely increase the amount of information available on this topic. To briefly summarise these findings, there is now significant evidence that religion or spirituality can help with coping with adversity (such as medical illnesses, adverse life situations and acts of terrorism), developing positive emotions (such as happiness, optimism, hope, meaning and purpose), managing depression, dealing with certain personality traits and possibly even reducing rates of suicide.

At a social level, there is evidence that religion and spirituality are associated with less delinquency and crime within communities, higher social support and greater social capital, which is a measure of community participation, trust and reciprocity. Overall, there is now a mountain of evidence to support the idea that religion and spirituality have the potential to boost individual happiness and well-being in all kinds of different ways.

Many more recent studies support this conclusion. To take just a handful of examples, a study from Qatar found that adolescents who consider themselves religious are happier, healthier and more satisfied with their lives.[44] Charismatic belief is associated with happiness among married women in rural Ghana[45] and there is a significant association between religiosity and happiness among university students in Turkey.[46] The overall evidence now strongly indicates that religiosity is a proven technique to attain happiness regardless of gender, nationality, race or religion.[47] So end of story? Religion boosts happiness, plain and simple?

While the research on this topic is consistent, it is worth noting one or two additional points. First, the history of religions does not always support the idea that religion makes people happy. Religion can be associated with intolerance, exclusion and even violence. While these are most commonly distortions of core religious tenets (virtually all religions preach peace), it is nonetheless the case that religious and spiritual ideas can be misused and thus increase unhappiness.

Second, David Myers and Ed Diener, two of the world's leading happiness researchers, add an important caveat to the recent literature on this topic.[48] They note that while there are demonstrated associations between religion and *individual* happiness, the position regarding religion and *communities* is less clear. Recent research findings indicate something of a paradox: while religious *individuals* in the United States report greater well-being, secular *countries* appear to have greater well-being than religious *countries*; while religious engagement is associated with greater life expectancy among *individuals*, it is associated with shorter life expectancy at *state level*; and while religious engagement is associated with lower crime rates among *individuals*, it is linked with higher crime rates at *state level*.

What these findings mean is that while it is clear that religion can increase an *individual*'s happiness, it is not necessarily the case that a community with many religious people in it is a happier *community*. While this might reflect difficulties with religious tolerance in certain communities, it might also reflect a need for more multilevel studies that examine possible tensions between religion at the individual level and religion at the societal level. Only time will tell.

Overall, current research strongly supports the idea that religious or spiritual belief and practice are beneficial for individual happiness, even if the position at community level is less clear. There is not yet sufficient research data to comment definitively on any possible differences between people who are religious and those who are 'spiritual but not religious', but it seems likely that at least some of the benefits of religion spill over into the latter group.

So if religious practices and beliefs can increase happiness, can political beliefs do likewise? The position with politics is more nuanced than with religion, but is equally fascinating.

The headline finding is that political outlook is linked with happiness and that conservatives consistently rate themselves as happier than liberals rate themselves. Much research now supports this position, confirming that a right-wing (conservative) orientation rather than a left-wing (liberal) one is associated with greater subjective well-being.[49] So, conservatives are happier than liberals.

How large is this effect? In 2006 one extensive telephone survey of 3,014 Americans found that 45 per cent of Republicans report being very happy, compared to 30 per cent of Democrats and 29 per cent of independents.[50] While not all Republicans are conservative and not all Democrats are liberal, there is a clear trend here: conservatism is associated with happiness. Why?

Various studies provide tantalising hints as to the reasons why this is. In 2015 one research group reported that conservatives have lower levels of neuroticism than liberals and this might account for higher life satisfaction among conservatives.[51] Another research group found that conservatives reported greater meaning and purpose in life than liberals did,[52] and yet another reported that conservatives express greater personal agency (i.e. control, responsibility), more positive outlook (i.e. optimism), more transcendent moral beliefs (i.e. religiosity, moral clarity) and general belief in fairness, compared to liberals.[53]

There is, then, no shortage of possible explanations for why conservatives are happier than liberals and no shortage of discussion about the implications of this kind of research,[54] although this pattern is not universal: there is some variation.[55] Even so, the overwhelming evidence shows that a person's political stance is linked with their happiness and that conservatives consistently rate themselves as happier than liberals.

It is, of course, entirely possible that this relationship works both ways: happiness or satisfaction with the world leads to conservatism, which boosts happiness. Conversely, unhappiness or dissatisfaction with the established order of things might be a requirement for certain forms of liberalism, which is often associated with a desire to change the world. Whatever way the relationship works, one thing is clear: conservatives are happier.

So far we have focused on the relationships between happiness and certain aspects of what we do with our lives: starting a family, earning money, gaining or losing employment, looking after our physical health and holding particular religious beliefs or political positions. Findings from research on these topics can help us to recognise both the importance and the limits of our life choices in shaping our happiness and well-being.

The happiness science presented here proves that what we do with our lives matters a great deal to our happiness, just as the previous chapter demonstrated links between happiness and who we are: our age, genetic inheritance, childhood experiences and place of residence. The next step is to move forward with this knowledge and progress into the areas of well-being, psychology and spirituality as applied in all our lives.

THREE

# Six Principles of a Happy Life

How can we achieve happiness? The ancient philosophers were full of thoughts on the subject. According to Aristotle, 'happiness is the meaning and the purpose of life, the whole aim and end of human existence'. Socrates advised that 'the secret of happiness is not found in seeking more, but in developing the capacity to enjoy less'. Seneca agreed, arguing that 'true happiness is to enjoy the present. A wise man is content with his lot, whatever it may be, without wishing for what he has not.'

But if acquiring material goods does not increase happiness, what does? What are the principles of a happy life? Is there enough commonality between people to allow us to sketch out some general ideas about how to achieve happiness? Does the science help?

The first two chapters of this book looked at findings from happiness research over the past few decades and suggested that specific patterns are identifiable in the distribution of happiness across populations.

Some of the research findings point to various ways forward. We should earn some money, but we should not depend on money or children for happiness. Both money and children can help with well-being, but neither is a magic ticket to eternal joy. We should prioritise employment, look after our physical health and pay attention to the roles of religious and political beliefs in our lives. We should strive to ensure that our children have happy childhoods and that our places of residence are conducive to well-being. If all else fails, there is the consolation of a general increase in happiness in later life, once we take care of our physical and mental health in the meantime. We could also move to Finland, or maybe Denmark, if nothing else works. From personal experience, I can certainly recommend Copenhagen.

These research findings are relatively clear and consistent, but where do they get us? What should we actually *do*?

This chapter moves on from the research and progresses into the areas of well-being, psychology and spirituality as applied in people's lives. How does all this work out in practice? The purpose of this chapter is to create a sustainable mental foundation for the second part of this book, which focuses on practical strategies to increase happiness in our day-to-day lives. This chapter takes a more general approach to the issue and presents the six overarching principles for a happy life that we can usefully keep in the back of our mind as we strive for greater well-being.

These principle are: (1) seeking balance (trying to achieve moderation in all things); (2) focusing on love (for ourselves and others);

(3) deepening acceptance (accepting what we cannot change and changing what we can); (4) practising gratitude (starting with the realisation that we are lucky to be alive); (5) avoiding comparisons with other people (which are the root of most human unhappiness); and (6) believing in something that matters to you, be it politics, religion, philosophy, football or even the emotional lives of minor celebrities. It does not really matter what you believe in, as long as you believe in *something*.

These six principles underpin the habits outlined and changes proposed in the next section of this book. Only by grappling with the philosophy can we successfully put these skills into practice – and keep practising them. I have selected these six principles out of an essentially infinite field of possibilities, based on the happiness science mentioned in the previous two chapters; common problems and solutions I see in my clinical work; and my own experience of studying happiness, well-being and mental health. These six principles also accord with the values of various philosophies and religions. They provide a sturdy but flexible framework for the changes that many of us need to make in our lives to achieve greater well-being.

Let's start with seeking balance.

## Balance

Lao Tzu was a Chinese philosopher who is said to have lived in the fourth or sixth century BCE and wrote a book called the *Tao Te Ching*. It is not clear whether Lao Tzu really existed or not. It is entirely possible that the figure of Lao Tzu is a composite one that emerged over the course of several centuries and eventually became attached to the series of philosophical sayings that is now presented as the *Tao Te Ching*.[1]

Whatever the truth might be, the idea emerged that someone called Lao Tzu wrote the *Tao Te Ching*, which is now firmly established as a foundational text in Taoism. Taoism is a philosophical tradition that advises us about how to live in harmony with the Tao, which is the sum of all of the rhythms and patterns of the universe. The Tao is also known as the 'Way' and it encapsulates the essence of everything that exists. The Tao is nothing less than the defining principle that lies at the heart of the cosmos.

As a result of its ubiquity and essential nature, living in harmony with the Tao brings harmony and balance to human existence. In order to achieve this in our busy, day-to-day lives, Taoism prioritises spontaneity, simplicity, compassion, frugality and humility. All of these qualities increase harmony with the natural world and, as a result, generate greater happiness.

In my life, I find Taoism's emphasis on simplicity especially useful. Life becomes very cluttered very quickly, with tasks, commitments, worries and repetitive behaviours that serve little purpose apart from keeping us busy. Even physical objects add to the complexity of our lives and can weigh us down as we rush from one activity to the next. Balancing work with family life and other commitments is always a challenge. For me, simplicity is the key to balance, and balance lies at the heart of the Tao.

Many other aspects of the Taoist approach to life involve achieving a balance between seemingly opposite or contrary forces, which might, in fact, prove complementary. In the words of Lao Tzu: 'Countless words count less than the silent balance between yin and yang.'

Achieving harmony and balance can be difficult. This is especially true when we are tempted by material desires (for money, power or other things), distracted by continual communication (especially social media) or simply locked in behaviour patterns that lead to excess rather than moderation. These are, however, precisely the circumstances in which the principles of Taoism are especially useful in guiding us towards happiness.

To follow the path of Taoism, it is not necessary to completely renounce one's current way of life and plunge headlong into Taoist philosophy. We can usefully start in a more moderate fashion by familiarising ourselves with the nature of Taoism and keeping its central concepts in our thoughts as we go from day to day.

This is easier than it sounds and brings swift rewards. Key Taoist ideas help us to trust more in nature and our bodies (an essential task in our hyper-cognitive age), cultivate more yin (reflection) than yang (action), and focus on wisdom instead of continually accumulating facts.[2] These steps can help us to slow down, connect better with the world around us and attain greater balance in our daily lives.

It sounds paradoxical, but we also need to be absent in order to present. We are often so immersed in our personal concerns that we lose sight of the bigger picture. We forget the broader landscapes in which our all-consuming personal dramas play themselves out. Sometimes, we even forget that a broader landscape exists! But if we step back and view ourselves from afar, we better understand where we are and what we need to do next. Removing ourselves helps us to become more deeply aware of ourselves and our intentions, and connects us with the present moment.

These apparently simple Taoist techniques promote the development of equanimity, harmony, integration and balance – all of which help with happiness.[3] According to Lao Tzu, we should 'do the difficult things while they are easy and do the great things while they are small. A journey of a thousand miles must begin with a single step.' Small changes matter most, especially when we are starting on a journey. We all possess great power within ourselves, but commitment to a purpose in life is an important step towards realising that power.[4] This does not mean that one rejects the free-flowing nature of Taoism or the idea of carefree wandering, but rather that one seeks to hear the calling of something greater than oneself. Having a destination in life provides freedom from uncertainty and the opportunity to achieve unity with the Tao along the way.

Do these ancient ideas have any relevance to happiness in the modern world? Taoism's emphasis on connecting with nature was never more important than it is today, when so many human lives are divorced from the rhythms of the planet and cause untold damage to the environment. Taoism's emphasis on moderation is also consistent with happiness science. We should exercise, but not excessively. We should earn money, but not too much. We should work, but not obsess. Having children might boost happiness, but not inevitably and not for ever. Moderation is key in all we do. In addition, our happiness set point has a large genetic component, so we need to accept this natural reality and learn to live with it to a certain degree – another task that reflects the essence of the Tao.

Conformity with nature and acceptance of what life brings lie at the heart of Taoism.[5] When good events occur, one should not be overwhelmed. When bad news arrives, the same applies. Lao Tzu made this point strongly: 'Life is a series of natural and spontaneous changes. Don't resist them – that only creates sorrow. Let reality be reality. Let things flow naturally forward in whatever way they like.' This is how we achieve balance and harmony with the Tao.

Taoism means following the natural order of things and being at one with nature.[6] If you realise and accept the truth of the world (the Tao), happiness will follow. Simplicity is key. According to Taoism, when you follow the laws of nature, you desire for nothing. When you step away from desire, happiness is what is left. Happiness lies not in material gain or social advancement, but in oneness with nature, balance in life and harmony with the Tao.

So what does this mean in practical terms? What can we actually *do* with this knowledge to increase our happiness?

First, in order to deepen our appreciation of Taoist wisdom, we can bear the principles of Taoism in mind in our day-to-day lives: spontaneity, simplicity, compassion, frugality and humility. These values promote balance and harmony, even if we simply reflect on them briefly from time to time as we move from task to task during the day.

Second, we can spend more time close to nature, simply observing the natural world and *feeling* our place within it. This is both a humbling and an elevating experience: humbling because everyone is small in the forest, and elevating because once we are in the forest, we are part of the forest, which is magnificent and huge. Simply being in communion with nature in this way nourishes our minds, restores a sense of balance and connects us with the environment, the planet and the Tao.

Third, we can engage in Taoist meditation and move towards deeper contemplation of the themes that lie at the centre of Taoist philosophy and so cultivate further realisation of the Tao. In his excellent book *Taoist Meditation: Methods for Cultivating a Healthy Mind and Body*,[7] Thomas Cleary notes that Taoist meditation is focused on both physical and mental health, consistent with contemporary understandings of the continuity between body and mind. By understanding, accepting and living according to the Tao, we can achieve greater physical and mental well-being.

Taoism encapsulates the first of the six overarching principles of a happy life: seek balance in all that you do and try to achieve moderation in all things. To return to Lao Tzu: 'Fill your bowl to the brim and it will spill. Keep sharpening your knife and it will blunt.' Focusing on moderation can be challenging. We always want more than we have and we sometimes deliberately do things to excess. With this in mind, is there an argument to be made that too much moderation is itself immoderate? Oscar Wilde certainly thought so, arguing that 'moderation is a fatal thing. Nothing succeeds like excess.' Taoism disagrees. According to the Tao, balance, moderation and harmony lie at the heart of the universe, the heart of nature and the heart of a happy life. I agree.

## Love

Many centuries ago love of God, love of country and love of family gradually gave way to an attractive, idealised vision of romantic love between two individuals. While many people are to blame for this state of affairs, William Shakespeare must be one of the chief culprits: in *Romeo and Juliet* the latter delivers a dangerously inspiring vision of romantic passion from a balcony in Verona: 'My bounty is as boundless as the sea, / My love as deep; the more I give to thee, / The more I have, for both are infinite.' Her beau responds with similarly stirring stuff: 'O blessèd, blessèd night! I am afeard, / Being in night, all this is but a dream, / Too flattering sweet to be substantial.'

Despite the fiery passions in this and myriad other plays, poems and romantic songs, love is a much broader concept than these melodramatic outpourings suggest. As humans, we feel real, lasting love for family, friends, pets, countries, ideas and – ideally – ourselves. Romantic love might routinely steal the headlines and take all the best songs, but other forms of love can prove equally reliable and sustaining. Some are even essential for human life. Let's start with parental love.

The love of a parent or guardian is vital for the well-being of a child. More specifically, happiness research shows that consistent, positive parenting and love are required if a child is to become a well-adjusted, happy adult. This is not just an emotional truth, but a biological one too. Children, in turn, can make their parents happy, especially when they are babies and again when they are adults, if circumstances permit.

What is the biology underlying these connections? Matthew Lieberman, Professor at the Social Cognitive Neuroscience Laboratory in the University of California, Los Angeles, writes about these issues in *Social: Why Our Brains Are Wired to Connect.*[8] Lieberman points to evidence that young children who are separated from their parents for long periods are at increased risk of long-term behavioural and literacy problems. Childhood stresses of this type are linked to alterations in brain regions associated with self-regulation in social settings, causing further problems in later life.

The link between love and survival is equally striking in animals other than humans. Lieberman summarises research on rats who underwent surgery to disconnect parts of the brain involved in attachment behaviour. Mother rats who underwent the surgery lost the ability to become fully attached to their young. Under testing experimental conditions, most of their babies died as a result. Similarly, in humans, love and nurturing are essential for growth and survival, not just as babies, but also during childhood, adolescence and adulthood. How does this work at a biological level?

First, evolution. Lieberman points to evidence that humans evolved bigger brains in order to socialise better. Groups of around 150 people seem to be ideal, so it is no coincidence that this was the average size of human settlements for much of our history. All told, the advantage of forming social connections is sufficiently strong to have driven an increase in human brain size over the course of evolution and thus increase our sociability and ability to connect with others.

Second, the functional biology of the brain, insofar as it is known, strongly supports the idea of a social brain that is primed for connection and love. Love has significant consequences for health and well-being, and so richly merits biological research but, for something that everyone talks about all the time, love has commanded little attention in the scientific literature to date.[9] This is a pity, because existing research points to identifiable stages of love, starting with the beginning of love (which can be stressful), followed (ideally) by relaxation of the stress response, followed ultimately (and again, ideally) by a more peaceful, beneficial state of sustained bliss.[10]

These three stages are reflected in both the psychology and biology of the individual.[11] From a psychological perspective, this process involves a complex combination of hope, trust, belief, pleasure and reward, often associated with the limbic or emotional areas of the brain. Positive outcomes reinforce specific loving behaviours, which lead to further positive outcomes such as communication, connection, compassion and, hopefully, happiness.

From a biological perspective, a large number of brain chemicals are involved in these processes: oxytocin, vasopressin, dopamine and serotonin among many others. These substances play multiple roles in the human brain in relation to virtually all emotions, including love. Oxytocin, for example, is often hailed as the 'love hormone', but while it is certainly involved in love, it has many other roles too. The same applies to serotonin, for example, which is often associated with happiness. All these substances have multiple functions and act in complicated, interacting ways to produce our thoughts, feelings and behaviours, including love.

It is worth remembering that, despite impressive advances in neuroscience, our knowledge of the human brain is still rudimentary. The brain is made up of some 86 billion nerve cells and a roughly equal number of other cells serving different, supporting functions. Brain chemicals convey messages between these cells and each nerve cell has approximately 7,000 connections with other cells. These connections are known as synapses and there are up to 500 trillion of them in the adult human brain. This means that our brains are, quite literally, mind-bogglingly complex. While we have some knowledge about how certain aspects of this extraordinary creation work, we still have only the dimmest ideas about the fine-grained chemical underpinnings of complex emotions like love.

Therefore, while we know that oxytocin and a range of other hormones and chemicals are involved in both social connection and love, the precise workings of love at a biological level remain obscure. This might always be the case. It is perfectly possible that our brains are too complex to understand themselves. Or, in the words of physicist Emerson M. Pugh, 'if the human brain were so simple that we could understand it, we would be so simple that we couldn't'.

We do, however, know that there *are* biological correlates of love, even if our knowledge of the brain is as yet inadequate to describe them in much detail. That is, perhaps, why we use the language of psychology or, indeed, burst into song or poetry when we talk about love: the language of biology is still not up to the task.

We do know that people in the early stages of a relationship have higher levels of the stress hormone cortisol and, in the longer term, love has the capacity to influence the limbic system, which is associated with emotions and reward pathways in our brains.[12] As a result, while the poetic essence of love might lie in our hearts and souls, the building blocks lie in the chemicals in our brains.

The end result of all of this discussion is that love, from one perspective at least, is a biological phenomenon that can bring great happiness, but, as a result, can also cause great suffering, both psychologically and biologically. Writing in *The Australian* in 2015, Ruth Ostrow, as a new mother, described feeling physical pain when she was not close to her baby.[13] This is an experience that many new parents will readily recognise and this pain is very real. As Lieberman points out in his book, it can be impossible to differentiate between social pain and physical pain on a brain scan. When we suffer, we suffer. The pain resulting from disturbances to social connection and love is essentially indistinguishable from the pain of a broken arm or leg. We are built to connect. Disconnection hurts.

Overall, happiness is the net result of love. This applies to all kinds of love: romantic love (when things are going well), parental love (which is essential for the well-being of children and the happiness of parents and guardians), love for others (our brains are evolved for social connection), love for the planet (which we have neglected) and love for specific activities that we enjoy doing and that increase our well-being.

So what can we do with this knowledge? First, we need to cultivate self-compassion in order to bring the benefits of love to bear on our relationship with ourselves. Too often, we find compassion for others but not ourselves. This is especially important in a book like this that is filled with advice about improving our well-being. It is vital that we have sufficient self-compassion to forgive ourselves when we fail to live up to our ideals and that we value our small achievements as we seek to increase our well-being. Change takes time. It is incremental. We should go easy on ourselves.

Second, love of others: the word 'love' is, perhaps, excessive for certain situations, but it is difficult to build any meaningful relationship that is not rooted in a solid foundation of love, fondness or (at the very least) positive regard for other people. Love is central.

Third, identifying activities that we love and doing more of them can boost our happiness in powerful, lasting ways. Perhaps the most intriguing research on 'happy actions' is a 2015 survey at the universities of Sheffield and Sussex, which looked at the relationship between daydreaming and well-being in 101 people.[14] They found that daydreams about significant others are associated with happiness, love and connection, but other, non-social daydreams are not. Even daydreaming about connecting with valued people increases feelings of love and happiness in our brains and (why not?) our hearts.

Perhaps, in the end, Shakespeare was right: love indeed holds the key to happiness, albeit that romantic love is not the only kind of love and moderation remains important. Even *Romeo and Juliet* points to the potential perils of over-exuberant romantic love when Friar Lawrence speaks like a Taoist as he warns of the danger of excessive passion and extols the path of moderation:

> 'These violent delights have violent ends,
> And in their triumph die; like fire and powder,

Which, as they kiss, consume: the sweetest honey
Is loathsome in his own deliciousness,
And in the taste confounds the appetite:
Therefore love moderately: long love doth so.'

## Acceptance

In addition to seeking balance and finding love, acceptance is the third overarching principle of a happy life. It is based on the idea that 'what you get is what you get' and it applies to some, but not all, aspects of our lives. The idea here is that excessive struggle against the inevitable not only is futile, but also distracts us from other areas of our lives in which we have more control. On this basis, we need to accept the things we cannot change and commit to changing what we can in order to achieve greater realism, balance and well-being.

This old idea has reappeared in multiple forms over several centuries. Perhaps the most familiar formulation is the Serenity Prayer, composed by American theologian Reinhold Niebuhr in the early 1930s and now associated with Alcoholics Anonymous: 'God, grant me the serenity to accept the things I cannot change, courage to change the things I can, and wisdom to know the difference.'

This idea of selective acceptance of the slings and arrows of life has many ancient echoes and is, perhaps, most linked with the Greek philosophy of Stoicism, dating from the third century BCE. Stoicism, however, incorporates much more than simple resignation to all the problems that life sends. More broadly, Stoicism argues that happiness lies in accepting each moment as it presents itself and not allowing fear of pain or desire for pleasure to control one's thoughts or actions. We should use our minds to understand the world, accord better with nature, cooperate with other people and treat others with fairness and justice.

To achieve all of this, it is important that we keep going, regardless of what life throws in our direction. Seneca, a famous Stoic, advised that 'even if some obstacle comes on the scene, its appearance is only to be compared to that of clouds, which drift in front of the sun without ever defeating its light'. Epictetus, another Stoic, regarded life's challenges as essential steps towards self-actualisation:

> What would have become of Hercules, do you think, if there had been no lion, hydra, stag or boar – and no savage criminals to rid the world of? What would he have done in the absence of such challenges? Obviously, he would have just rolled over in bed and gone back to sleep. So, by snoring his life away in luxury and comfort he never would have developed into the mighty Hercules.

Seneca agreed that overcoming obstacles is essential for well-being: 'No man is more unhappy than he who never faces adversity. For he is not permitted to prove himself.' Against this background, Stoicism recommends that we overcome unhelpful emotions by developing self-control and fortitude, and that we improve the clarity of our thought by removing bias. 'Virtue is the only good,' according to the Stoics, and external things, such as wealth and pleasure, are just 'material for virtue to act upon'.

The idea of accepting the present moment as it is can be found in many philosophical and spiritual traditions, not just Stoicism. This idea has also contributed to the development of a relatively new form of psychological therapy, known as 'acceptance and commitment therapy'. This therapy is rooted in a combination of acceptance, mindfulness and behaviour change, aimed at increasing psychological flexibility. Developed by Steven Hayes in 1982, acceptance and commitment therapy integrates cognitive and behavioural approaches in order to create a unified, process-based model of how to alleviate psychological problems and advance behavioural effectiveness. This description is quite the mouthful. Does this therapy work? And how, exactly, is it carried out?

By June 2019 there were over 280 randomised, controlled trials of acceptance and commitment therapy, involving a total of almost 33,000 participants.[15] Outcomes are reported as being as good as, or in some cases better than, other psychological therapies.

So how does acceptance and commitment therapy work in practice? There are six core components of the therapy: (1) acceptance or non-judgemental awareness of thoughts, feelings and sensations; (2) de-fusion or shifting focus from the content of thought to the process of thinking; (3) mindfulness of the present moment (which features heavily in Buddhism too); (4) moving from seeing the 'self' as content to seeing the 'self' as context (not unlike seeing oneself from afar in Taoism); (5) clarifying valued directions and actions (knowing where you want to go); and (6) committing to actions that lead in valued directions, despite the presence of unwanted feelings or thoughts.[16]

The specific actions taken by the therapist and client in this therapy can involve identifying metaphors for unworkable behaviours (e.g. 'tugs-of-war'), letting go of specific behaviours, writing thoughts on cards, focusing and refocusing mindfully on the moment, stepping back to see one's actions from a new perspective, describing personal values as 'directions' and then building new patterns of action with small steps that lead in valued directions. The key to much of this lies in feeling feelings as feelings, thinking thoughts as thoughts, and then getting on with the business of living.[17]

Acceptance is an important first step in this process and remains central to sustaining and enhancing the changes required for effective living. As a result, acceptance and commitment therapy is one of the most important developments in clinical psychology in recent decades. Its ideas are not only useful in the context of therapy itself, but can also be applied in smaller ways in everyday life. It is particularly important that acceptance is always coupled with commitment to change, especially changes in behaviour. We cannot always 'think our way' out of negative thought patterns. Sometimes, behavioural changes are the most effective way forward.

Clinical psychologist Tara Brach focuses specifically on the role of acceptance in generating positive change in her book, *Radical Acceptance: Embracing Your Life With the Heart of a Buddha.*[18] Brach draws artfully on Buddhist philosophy and psychology to argue that accepting ourselves and our lives as they are is a necessary and even urgent antidote to our tendencies to neglect ourselves, pass negative judgements and generally treat ourselves harshly. Radical acceptance means experiencing ourselves and our lives exactly as they are, without judgement or comment. This, Brach contends, leads to freedom.

Even though Brach is clearly right, most of us find that overcoming feelings of unworthiness is not easy, especially when our negative thinking habits were laid down many years ago and are now deeply ingrained. But positive change is always possible and Brach outlines both case histories and techniques to provide us with the necessary guidance. Meditation, for example, can help prevent us remaining lost in our thoughts, especially the recurring negative ones. Learning to pause in our busy lives is also useful, as is connecting with the present moment and practising how to truly inhabit our bodies.

In my life I have found that meditation is a useful way to sit with the present moment and accept it as it is.[19] I discuss meditation again in later chapters and also in *The Doctor Who Sat For a Year*, a book about a year spent trying to meditate every day.

Meditation helps not only with acceptance, balance and calmness, but also with compassion and connection. All of these concepts are key elements of Buddhist practice and they help us to accept ourselves more deeply. Actively developing compassion is another vital part of this process, once we awaken compassion not only for others but also for ourselves. Compassion is both a value and a skill. Working on compassion for ourselves helps us to recognise our basic goodness and to accept both who we are and what we do with our lives.

The path outlined by Brach involves a subtle mixture of acceptance and change, just as acceptance and commitment therapy combines certain forms of acceptance with clear steps towards better understanding and more effective living. Acceptance and change are essential in any process of self-renewal. Acceptance without change is stagnant, and change without acceptance is futile. A careful combination of focused acceptance and mindful change can, however, increase happiness by both acknowledging everything that life sends our way and recognising our agency and power to produce radical shifts in our lives and worlds.

As the Serenity Prayer advises, we need to accept what we cannot change and find the courage to change what we can. Challenges and impediments are an inevitable part of life, but many obstacles disable us only if we let them. By accepting and navigating life's challenges, we learn to flourish and grow as people. As Roman Stoic Marcus Aurelius argued, 'the impediment to action advances action.

What stands in the way becomes the way.' Accepting this fact can be difficult, but it does help us live happier lives.

## Gratitude

Gratitude is having a moment. The past decade has seen an avalanche of books about gratitude, thankfulness diaries and gratitude courses in schools. Most of these initiatives focus on recognising the many good things in our lives, giving thanks for what we have, and sharing our resultant gratitude with other people (to whom we should feel grateful for being in our lives in the first place). We are advised to want what we already have, rather than spend our lives wishing for things that we think would make us happy (but probably wouldn't). In truth, it's all a bit exhausting. Perhaps I have gratitude fatigue?

The gratitude movement is certainly welcome and warming, but does it make us happier? Do we really enjoy greater well-being if we coax ourselves to rustle up gratitude for things we already have? And, if we do benefit from practising gratitude, is it the gratitude itself that helps or is it the underlying impulse that made us buy that gratitude diary in the first place? Which comes first: the gratitude or the happiness?

The first point to make is that humans have always advised each other to be more grateful. This is one of our favourite pieces of advice of all times. It stems from both our general tendency to give each other unsolicited advice and the belief among advice-givers that advice-receivers do not fully appreciate their good fortune (unlike the advice-givers, who presumably do). As a case in point, Epictetus, a Stoic, held that 'he is a wise man who does not grieve for the things which he has not, but rejoices for those which he has'. English writer

G.K. Chesterton maintained 'that thanks are the highest form of thought; and that gratitude is happiness doubled by wonder'. The gratitude industry is based on this idea, allied to the belief that being grateful makes us happy. But does it?

In 2019 researchers in Brazil published a relatively large study of positive psychology and gratitude interventions.[20] These researchers assigned 1,337 participants to one of three groups. Each evening, the 'gratitude group' wrote down five things for which they were grateful that day; the 'neutral group' wrote down five things that affected them in any way (positive, neutral or negative); and the 'hassles group' wrote down five things that annoyed them. Fourteen days later, researchers found that the gratitude intervention increased positive affect, subjective happiness and life satisfaction, and reduced negative affect and depressive symptoms. There was some similarity between the 'gratitude group' and the 'neutral group', but this appears to be because the neutral group tended to write down positive events rather than negative ones on their lists.

These findings provide strong evidence that actively cultivating gratitude makes us happier and less depressed. But this is just one research study. Do other studies produce the same result?

A different group of scientists performed a meta-analysis of research in this area in 2016, looking at results across 26 studies of gratitude interventions.[21] Results confirmed that gratitude interventions boost psychological well-being significantly more than other interventions or no interventions. They conclude that longer interventions and more powerful studies are needed, but that results up to 2016 support the idea that gratitude interventions show positive promise. While the value of such interventions might vary across cultures,[22] it is likely that cultivation of greater gratitude would benefit most people. Three years later, the work from Brazil confirmed the value of gratitude practices for boosting well-being and reducing feelings of depression.

Particular groups might benefit more than others. A 2014/15 study indicated that there is an inverse relationship between gratitude and loneliness; i.e. the more gratitude a person experiences in daily life, the less lonely they are.[23] In fact, variations in gratitude between individuals account for up to almost one-fifth of the variability in loneliness, reflecting the key role of gratitude in determining how lonely a person feels. Other risk factors for loneliness in this study included being female, not having a stable relationship and not participating in the workforce. While it can be difficult to figure out if loneliness impacts on gratitude, or if gratitude impacts on loneliness, there is no doubt that lack of gratitude is linked with loneliness, and that interventions to increase gratitude might well help diminish loneliness.

Overall, then, what does research about gratitude tell us about living a happy life? How grateful should we aim to be?

Plainly, it is important that we are grateful for what we have, but we must also be realistic, or else gratitude loses its meaning and value. We should not spend every day trying desperately to rustle up gratitude for things that we do not really feel thankful for. On the other hand, we should be genuinely grateful for some of the enormous gifts that we routinely overlook. With this in mind, there are four key messages about gratitude as an overarching principle for a happy life.

First, we should be grateful that we are alive. The fact that we are living and able to read this book is partly a matter of skill (we have successfully eluded death so far), but it is mainly a matter of luck.

Most people are dead, and we are not. Statistically, we are incredibly, ridiculously, outrageously lucky to be alive. Around 107 billion people have ever lived on earth and we are among just 7.6 billion currently living.[24] The fact that this good fortune is largely due to chance is all the more reason to cherish it and make the most of our lives. Sometimes, this means being happy. Sometimes, it means just keeping going. Always, it means being grateful.

Second, we should be grateful for other people. While we tend to attribute most of our woes to other people (as opposed to ourselves), we should remember that they (like us) are just trying to be happy. And, like us, there is nothing that other people value more than being valued. American psychologist William James pointed out that 'the deepest principle in human nature is the craving to be appreciated'. This applies to other people just as it applies to us. Being grateful to others is not just an honest acknowledgement of all that other people bring to our lives, it is also smart people management: everyone loves to be loved.

Third, simply expressing gratitude is not enough. Recording thankful thoughts in a diary or journal will deliver benefits, but taking grateful actions (doing someone a favour, giving a gift, etc.) provides a much bigger boost to well-being. President John F. Kennedy said that 'as we express our gratitude, we must never forget that the highest appreciation is not to utter words but to live by them'. Psychologist Maureen Gaffney outlines useful gratitude practices along these lines in her book *Flourishing: How to achieve a deeper sense of well-being, meaning and purpose – even when facing adversity*.[25] Gaffney suggests developing a habit of being grateful by recording three good things that happen to you each week. This simple practice can help improve your mood within a matter of weeks, especially if you translate your grateful feelings into grateful actions.

Fourth, we should be grateful for having enough in our lives, as opposed to wishing we had more. Buddhism is particularly strong on the idea that enough is enough. There is even a Buddhist proverb: 'Enough is a feast.' Accumulating too many possessions not only fails to make us happy, it makes us unhappy. This is deeply regrettable, especially when there is always something to be thankful for, once we reframe our situation properly. In the words of one Buddhist teacher: 'Let's rise and be thankful, for if we didn't learn a lot today, at least we may have learned a little. And if we didn't learn even a little, at least we didn't get sick. And if we did get sick, at least we didn't die. So let us all be thankful.'[26]

## Avoiding comparisons

As humans, we appear to be programmed to compare ourselves with other people. This habit persists despite the fact that it clearly makes us unhappy. We have been advised against social comparisons for many centuries, with little effect, if any. The Bible takes a firm line in the Book of Corinthians: 'Not that we dare to classify or compare ourselves with some of those who are commending themselves. But when they measure themselves by one another and compare themselves with one another, they are without understanding.'

Similar advice is provided in Islam and throughout Buddhism, which suggests that you 'focus, not on the rudenesses of others, not on what they've done or left undone, but on what you have and haven't done yourself'. In essence, the great majority of religious and philosophical traditions agree that we should not compare ourselves with other people, but instead focus on our own thoughts and behaviours.

Despite all of this excellent advice, we continue to compare ourselves with other people multiple times a day, in conversation, on social media or in a million other ways. An enormous mountain of human unhappiness stems from this habit, even though such comparisons are never accurate, are rarely informative and inevitably make us depressed. At an intellectual level, most of us already know this. So why do we persist with such self-destructive behaviour?

In 1954 American psychologist Leon Festinger presented 'a theory of social comparison processes'.[27] Festinger argued that humans have a drive to obtain accurate self-evaluations. More specifically, he hypothesised that people try to evaluate their opinions and abilities through objective, non-social means, but, if such means are unavailable (as they often are), we evaluate ourselves by comparing ourselves with other people. And while we try to compare ourselves with people who are somewhat like ourselves, most western societies place a premium on doing better and better, and this fundamentally affects the comparisons we choose to make.

The net result of this process of social comparison is often not the attainment of accurate self-evaluations, but the development of a compulsive cycle of comparison that makes us dissatisfied and unhappy. In recent years this process has been greatly amplified by social media, which permit rapid, vapid comparisons with all kinds of distant figures, despite the enormous differences between our situations and theirs. Most commonly, we compare ourselves on an average day with a celebrity whose online persona is the result of much careful preparation, hours of work by a team of stylists and immaculate online curation. Therefore, while social comparison is not new, it is faster and more potent in contemporary culture than it ever was before.

Comparison is a fundamental building block of human thinking and a basic element of human social life. It has been found that social comparisons are linked with cultural practices that promote strong norms and punishment for deviance (i.e. social tightness or conformity) and relational self-construal (i.e. collectivism or group membership).[28] While social comparison might be a fundamental human tendency, it is influenced by the social context in which it occurs, as we see with contemporary social media. The impulse to compare might be as old as humankind, but the ways we do it change with time.

So is there a clear link between social comparison and unhappiness, and, if so, what can we do to alter our habits and increase our well-being?

Richard Easterlin argues that people devote an excess of time and attention to pecuniary or financial objectives (such as work, wealth and possessions) as opposed to non-pecuniary ones (such as health and family life).[29] In relation to social comparison, he suggests that theories of happiness should be built on evidence that social comparison affects utility or outcomes more in pecuniary than non-pecuniary domains. Social comparisons in the areas of wealth and finance are more accessible and corrosive than comparisons in relation to health or family.

What this means is that we can readily compare our wealth with that of our neighbour (and this makes us unhappy), but we can less readily compare our health or family with those of our neighbour, with the result that we do less comparison across the domains of health and family – and our happiness is less affected as a result. On this basis, Easterlin recommends the development of policies that reflect these findings and place greater emphasis on non-pecuniary aspects of life, such as health and family, that are both relatively neglected and less prone to unhelpful social comparison. This makes perfect sense: the less comparing we do, the happier we will be.

What policies might help us to achieve this outcome and reduce the negative effects of social comparison?

The 2020 *World Happiness Report* provides valuable information about this in a chapter titled 'The Nordic Exceptionalism: What Explains Why the Nordic Countries are Constantly Among the Happiest in the World'.[30] The authors note the relationship between social comparison and well-being, and point out that people's subjective perception of their position in society is more predictive of their well-being than measures such as income. They also report that this effect is less pronounced in Nordic countries that have strong welfare states compared to other countries. Being poor in Denmark, for example, does not have the same impact on happiness as being poor in the United States. As a result, social comparison does not carry the same weight in Nordic countries as it does in the United States and many other nations.

The conclusions of both Easterlin and the 2020 *World Happiness Report* point governments and policymakers strongly in the direction of policies that prioritise health and family values over financial success and strengthen welfare provisions. In addition to these high-level steps, however, there is also much that we can do as individuals to minimise the negative effects of social comparison in our lives.

In 2016 some five studies were conducted to see if seeking companionship as opposed to comparison distinguished happy people from unhappy people.[31] In three studies, participants were put in a hypothetical social situation requiring them to place more value on either companionship or comparison. That is, participants had to place a value on a social interaction that involved companionship but was threatening in terms of comparison (e.g. meet a happy but apparently more competent or talented friend) and another social interaction that involved comparison but less companionship (e.g. meet an unhappy but apparently less competent or talented friend). This was a difficult choice.

Intriguingly, the researchers found that happy people predicted that spending time with a happy, 'superior' friend would be more enjoyable than spending time with an unhappy, 'inferior' friend. Happy people prioritise social companionship over social comparison. Happy people were also more willing to socialise with a happy, superior friend than with an unhappy, inferior friend, further reinforcing the researchers' point that happiness is associated with valuing social companionship and resisting the siren call of social comparison. The researchers performed two other studies to further clarify their findings and reported that their results were not explained by self-esteem or by happy people simply being attracted to other happy people. The final outcome was clear: happiness is linked with companionship rather than comparison.

Overall, this research points to the first step that we can take as individuals to minimise the negative effects of social comparison: we need to actively seek out companionship (which makes us happy) rather than comparison (which doesn't). This is an important first step, not only because it reduces the time we spend comparing, but also because companionship is a healthy emotional substitute for comparison. What else can we do?

In 2018 Susan Biali Haas, an award-winning physician who speaks and writes about stress management and mental health, addressed this issue in her splendid blog on the *Psychology Today* website.[32] Biali Haas starts by noting President Theodore Roosevelt's comment that 'comparison is the thief of joy'. Roosevelt was absolutely right and Biali Haas points out that the tendency to compare ourselves with others is as old as humanity itself. She highlights the role of social media in this area and writes that she tries to use social media purposefully, choosing carefully what she looks at.

In terms of advice, Biali Haas suggests that we develop awareness of how we might inadvertently provoke comparison in others and how we get sucked into comparison ourselves. She advises us to identify and avoid our comparison triggers, cultivate gratitude in our lives, and remind ourselves that we are not comparing like with like when we compare our life today with someone else's profile on social media. Biali Haas's advice is excellent. Remember: you have no idea what other people's lives are really like. Just live your own.

In the final analysis, the key to avoiding comparisons with other people is to focus on what is 'enough' for you and not what you achieve compared to other people. Such comparisons are inaccurate, unhelpful and unwise. In the words of Lao Tzu: 'When you are content to be simply yourself and don't compare or compete, everybody will respect you.' Marcus Aurelius agrees: 'How much time he gains who does not look to see what his neighbour says or does or thinks, but only at what he does himself.'

This path is difficult but rewarding, and it is the fifth overarching principle for a happy life: insofar as humanly possible, avoid comparing yourself with other people.

## Believing

Having discussed five overarching principles of a happy life – balance, love, acceptance, gratitude and avoiding comparisons – we come to the sixth: believing. This is a tricky one.

There is evidence that religion and spirituality contribute to individual happiness and good mental health. Despite the associations between certain religions and concepts such as guilt and intolerance, the net effect of religion and spirituality is positive for most people. There is also evidence that a person's political stance is significantly related to their happiness, with conservatives consistently reporting themselves as happier than liberals. Religion and politics touch on the issue of belief in different ways: religion is especially interesting.

I have vivid recollections of a discussion I had with a Roman Catholic nun when I was a teenager. For some reason that escapes me now, we were having a detailed debate about the nature of religious faith. I told the nun that there was no hard evidence that God exists and therefore no reason to believe. The patient nun paused before responding: 'If there was hard evidence, belief in God would be a matter of science. Nobody is saying that. Belief in God is a matter of faith. Faith is a different kind of belief.'

This conversation stayed with me for two reasons. On the one hand, I realised that I had fallen into a thinking error generated by two ostensibly reasonable propositions. The first proposition is that things that are supported by evidence are true and the second proposition is that things that are not supported by evidence are untrue. As the years rolled by, it became clear to me that both propositions are shaky. The first proposition is only as reliable as the evidence available (which is rarely definitive) and the second is simply foolish: absence of evidence is not evidence of absence.

The conversation with the nun also stayed in my mind for several decades because it taught me that science provides excellent reasons to believe or disbelieve certain things, but, where scientific evidence is insufficient or impossible, many people *choose to believe certain things anyway.* This is not necessarily an error: we mostly know when our beliefs are based on scientific evidence and when we choose to believe. The latter is called faith, and the more I understand the limitations of science, the more I admire people of faith.

This does not mean that we should just believe anything that comes along simply because we want to believe it. Science provides plenty of certainty about some matters and pretty reliable hunches about others. No, it means recognising the limits of factual knowledge and clearly labelling the occasions when, in a field of uncertainty, we *choose* to believe something. While some religious people regard the existence of God as a rock-solid truth, many admit that it is a matter of faith; something that they *choose* to believe. This is an important distinction.

Political belief is slightly different, but the fundamental principles are strikingly similar to religious faith. Evidence about specific politicians, political parties or ideologies generally accounts for just some of the reasons why we adopt particular political beliefs. We also have many non-evidence-based reasons for our political choices: family traditions, idiosyncratic preferences or a simple desire to believe in *something*, leading to essentially random decisions about election candidates or political affiliations. All of these factors can be relevant and all indicate a desire to believe in something rather than nothing, even when our beliefs stray well beyond the limits of the evidence available.

The human search for meaning lies at the core of this hunger for belief. Viktor Frankl, the Austrian neurologist, psychiatrist and

Holocaust survivor wrote movingly about this in *Man's Search for Meaning*, his 1946 book about his experiences as a prisoner in Nazi concentration camps during World War II.[33] Even in situations of unthinkable suffering, meaning sustains the soul.

More recently, Emily Esfahani Smith highlights the role of meaning in *The Power of Meaning: The True Route to Happiness*.[34] Smith argues that a search for meaning can be more fulfilling than an explicit search for personal happiness. She also writes – convincingly – that contributing to something that is bigger than each of us helps us to attain greater well-being in our lives. The *Bhagavad Gita*, a key Hindu scripture, goes even further, holding that belief is what defines us as people: 'Man is made by his belief. As he believes, so he is.'

Mahatma Gandhi, who used nonviolent resistance to lead the successful campaign for India's independence from Great Britain, had a profound understanding of the power of belief:

> Your beliefs become your thoughts,
> Your thoughts become your words,
> Your words become your actions,
> Your actions become your habits,
> Your habits become your values,
> Your values become your destiny.

Ultimately, what this means is that belief is uniquely conducive to human flourishing and well-being. In a sense, it does not really matter *what* you believe – as long as you *believe*.

These, then, are the six overarching principles of a happy life, informed by the findings of happiness research and combined with insights from the literatures on well-being, psychology and spirituality. The six principles are: balance, love, acceptance, gratitude, avoiding comparisons and believing. These principles appear again and again in the philosophies and religions of the world and are in clear evidence throughout the psychological literature and happiness science that we have discussed.

An engagement with these beliefs provides a firm foundation for the advice in part two of this book, which presents practical strategies to increase happiness in our day-to-day lives. What should we actually *do* in order to be happier?

PART TWO

# Strategies for Happiness

FOUR

# Sleep/Wake Up

I love to sleep but am not very good at it. I gaze with envy at the cat, Trixie, who sleeps for lengthy periods at any time of day or night, regardless of where she is. Trixie sleeps for ten hours overnight, wakes up slowly, enjoys a leisurely breakfast and then makes her way to someone's bed to sleep for another few hours. This is followed by a period of time outside, often spent asleep, after which she returns to the house for a nap. Trixie sleeps anywhere, regardless of the noise level or activity around her. If she is woken for any reason, she looks around, yawns and immediately returns to sleep. I have much to learn from that cat.

I am the opposite of Trixie. I sleep poorly. I have no problem getting to sleep, but I wake up early. If I am lucky, I sleep until 6 a.m., by which time I am always awake, regardless of what time I went to bed. Sometimes I wake at 4 a.m. and cannot get back to sleep. Around once a year, I sleep until 7 a.m. Those days are magical: I am full of zest and life. I need more of those days and fewer unsettled, broken nights.

For many years, I have been convinced that more, better sleep would make me happier. Macbeth would agree: here he describes the many benefits of a good night's rest: 'Sleep that knits up the raveled sleave of care, / The death of each day's life, sore labour's bath, / Balm of hurt minds, great nature's second course, / Chief nourisher in life's feast.'

How can I achieve this? How can I better nourish 'life's feast' with more sleep?

Good sleep is one of the most important and neglected ingredients of a happy life and we should all should spend more time asleep at night and more time awake during the day. This chapter advises about 'sleep hygiene' to improve sleep at night-time and focuses on the idea that we should 'awaken' more in our daily lives, as advised by virtually every philosophical and spiritual tradition on the planet.

Let's start with sleep.

## Why do we sleep?

Everyone sleeps. This fact alone is sufficient reason to conclude that sleep is vital for the well-being of humans and other animals. If we are deprived of sleep, we could die. These are two compelling arguments in favour of sleep: everyone sleeps and those who do not sleep could die.

Recent decades have seen scientists delve deeper into the reasons why sleep is so important for our well-being and what light the mysteries of sleep can shed on how our bodies function. The results of this research strongly underpin the value of sleeping, provide insights into the complexity of sleep itself and still leave much research to be done before we understand sleep fully. While sleep is slowly revealing its secrets, there is much yet to be discovered.

Even at this stage, however, we know that most of us need to get more sleep. In the United States, the National Sleep Foundation recommends that adults need 7 to 9 hours sleep in every 24 hours. In my experience, few people meet this requirement. I know I don't. The National Sleep Foundation recommendations were updated in 2015 to take account of the most recent research and are summarised in the table below..

| **How Much Sleep Do We Need?**[1] | |
|---|---|
| *Age group* | *Recommended sleep needed in each 24-hour period* |
| New-borns (0 to 3 months) | 14 to 17 hours |
| Infants (4 to 11 months) | 12 to 15 hours |
| Toddlers (1 to 2 years) | 11 to 14 hours |
| Pre-schoolers (3 to 5 years) | 10 to 13 hours |
| School-age children (6 to 13 years) | 9 to 11 hours |
| Teenagers (14 to 17 years) | 8 to 10 hours |
| Younger adults (18 to 25 years) | 7 to 9 hours |
| Adults (26 to 64 years) | 7 to 9 hours |
| Older adults (65 years plus) | 7 to 8 hours |

As these recommendations show, we need different amounts of sleep at different stages of our lives, with the number of hours that we require declining as we grow older. This pattern is clear across multiple research studies and prompts a fundamental question about the purpose of sleep: why do we need to sleep at all?

Sleep restores: our bodies do not 'switch off' when we fall asleep, but move into a different mode of activity. While we sleep our bodies perform many essential functions. Some of these functions are well recognised and understood, but many remain mysterious. Sleep has not yet yielded its most profound mysteries to science.

Matthew Walker provides an excellent overview of the uses and value of sleep in his much-celebrated book, *Why We Sleep: The New Science of Sleep and Dreams.*[2] Walker is Professor of Neuroscience and Psychology at the University of California, Berkeley, and a world expert on sleep. He points to many of the proven benefits of sleep, starting with learning. Sleep refreshes our ability to make new memories and, once we have learned something, sleep helps us to embed it in our minds. Therefore we should get a good night's sleep both before and after we learn new material. Interestingly, sleep also plays a vital role in forgetting or, rather, selecting which material to retain in our memories and which material to let go. Forgetting is essential for mental well-being, and sleep is essential for forgetting.

Walker also looks at the flipside of sleep: the effects of sleep deprivation. He presents compelling evidence linking sleep deprivation with emotional irrationality, characterised by swings to extremes of positive and negative emotions, as well as lapses in concentration that lead to road-traffic accidents. In the longer term, lack of sleep appears to increase the risk of Alzheimer's disease, a debilitating form of dementia that affects more than 40 million people around the world. Other health effects of a lack of sleep include increased heart rate, raised blood pressure, weight gain, heightened risk of diabetes and negative effects on our reproductive and immune systems.

All told, sleep is an essential element of virtually all body systems, ranging from our brains to our hearts, from weight regulation to reproduction. With this in mind, it is not surprising that there are also strong links between sleep and our emotional lives, especially emotional regulation and happiness.

Daytime events, especially emotionally stressful events, have noticeable impacts on both sleep quality and well-being.[3] Researchers have highlighted significant variations in activity in the brain's limbic system during sleep. The limbic system is involved in emotional modulation, so activity here during sleep likely indicates some form of emotional processing that is essential for well-being. There is an intimate relationship between our sleeping and waking lives: not only do daytime events influence sleep, but sleep, in turn, influences how we react to emotional and stressful stimuli during the day. Our sleeping hours and waking hours are closely linked with each other.

Cara Palmer and Candice Alfano at the University of Houston agree with this assessment and provide a valuable overview of published research concerning sleep and emotion regulation.[4] They point out that many of the same brain structures and chemicals are involved in both emotion regulation and sleep, with the result that the relationships between emotions and sleep are both intimate and complex. They draw a distinction between emotion generation, which concerns emotional responses that we experience, and emotion regulation, which concerns how we influence what emotions we have, and when and how we have them. To date, most research about sleep and emotions has focused on emotion generation, but there is a strong argument to be made that emotional regulation deserves more attention.

Palmer and Alfano conclude that inadequate sleep produces more negative emotions and fewer positive ones, and that sleep deprivation can have a negative impact on emotion at several stages of the regulatory process. In practical terms, this means that more and better sleep will likely increase our positive emotions, decrease negative ones and help us to better control and manage our emotional lives. This is vital insight: well-being is determined not only by which emotions we experience, but by how we manage those emotions. On occasion, we have limited control over which emotions appear in our lives, but we always have significant control over how we manage and regulate our emotional responses. Sleep is essential for all of these functions and deserves enhanced attention as a result.

Sleep also plays a vital role in encoding, consolidating and retrieving emotional memory traces.[5] Sleep deprivation alters emotional reactivity to both positive and negative events and reduces our ability to share in the emotional states of others: sleep deprivation diminishes empathy. The links between sleep and these essential emotional functions are increasingly underpinned by neuroscientific findings that consistently demonstrate a delicate, intricate network of brain activity while we sleep. The human brain is a magnificent creation that needs to sleep in order to perform its magic.

Vanessa King describes many of these benefits in *10 Keys to Happier Living: A Practical Handbook for Happiness*.[6] King recalls going to bed an hour or so earlier than usual while she was away on a break and then continuing the habit once she returned home. The effects were amazing: the extra hour of sleep improved King's outlook, improved her interactions with other people and improved her ability to deal with the events of each day. King concludes that sleep is not a luxury, but an essential for health and well-being. The biological evidence we have discussed so far in this chapter strongly supports her view.

In light of both King's view and the emerging science of sleep, it is not surprising that I feel so much better on those rare occasions when I wake at 7 a.m. as opposed to my usual 6 a.m. To borrow the words of Macbeth, 'life's feast' is better nourished by that extra hour of sleep and I feel infinitely more ready for the day after I have enjoyed my (relative) lie-in.

King also notes that a grateful state of mind and a focus on positive thoughts can help with restful sleep. As the Dalai Lama points out: 'Sleep is the best meditation.' Sleep is also the best way to restore our bodies, nourish our minds and prepare us for the day to come. All of this is now clear, especially in light of recent research about sleep, but there is still one more question to answered: does sleep actually make us *happy?*

## Does sleep make us happy?

As we saw in the opening chapters of this book, happiness is influenced by many factors: who we are, where we live, what we do and how we choose to live our lives. While no single factor can be singled out as having an especially decisive influence on well-being, the evidence discussed so far strongly suggests that good sleep can make a significant contribution to physical health and vitality. But is there evidence linking sleep with happiness in particular? Does more sleep mean more joy?

There is now compelling evidence that sleep is significantly associated with positive affect, which is the extent to which a person experiences positive mood, happiness, interest and joy.[7] But while good sleep is associated with greater happiness, it is not clear if sleep increases happiness, or if happiness promotes sleep, or indeed if there is a two-way relationship between sleep and happiness.

In all probability, sleep and happiness influence each other in different ways, with a good sleep pattern underpinning happiness during the day and happiness, in turn, supporting good sleep at night. Both the quality and quantity of sleep are likely to matter, because broken or disturbed sleep is unrefreshing regardless of how long a person stays in bed. The relationship between sleep and happiness also changes over time: our need for sleep generally diminishes as we grow older and various aspects of our changing lives have a knock-on effect on our physiological and psychological needs.

It is also possible that different age groups experience different relationships between sleep and happiness. For example, reduced sleep is associated with a 55 per cent increase in the likelihood of mood problems among adolescents including, ultimately, anger and depression.[8] This effect is consistent across geographical regions, leading to the conclusion that sleep is a universal and modifiable risk factor for preventing mood problems in this at-risk age group.

This is a finding with important public health implications. Good sleep hygiene might well decrease depression and even increase happiness among young people. We already know that people in this age group are at significant risk of anxiety and a range of other psychological and psychosocial problems. Enhancing sleep is a relatively non-conflictual way to promote well-being and hopefully decease negative emotions, enhance coping skills and boost happiness during this critical period in their lives.

All of these research findings will make perfect sense to anyone who has had problems with sleep in the past: sleep deficits have a clear and lasting negative effect on mood. Moreover, after a few good nights' sleep, the effect is instant. It is like magic. The world seems better, brighter and infinitely more manageable after a good night's sleep. Even other people seem more reasonable and problems that were insurmountable yesterday seem either solvable or less important in the morning.

Mental illness can affect both sleep and mood and, for people with depression, problems with sleep can be profound. Depression is characterised by a broad range of symptoms including low mood, loss of interest and enjoyment, reduced concentration and attention, diminished self-confidence and self-esteem, a pessimistic view of the future, hopelessness, helplessness and ideas or acts of self-harm.[9] The condition is also associated with a variety of physical symptoms including loss of appetite, reduced energy, marked tiredness after slight effort and, for many people, disturbed sleep. Sleep disturbance is one of the most distressing aspects of the condition.

Sleep problems in depression can include difficulty falling asleep, fragmented sleep, nightmares and early morning waking. Some people with depression experience hypersomnia, which is excessive sleep. Whatever the nature of the sleep problems in depression, they appear to be intrinsically linked with low mood and generally improve when depression responds to treatment. This invariably comes as an enormous relief because most of us do not fully appreciate the restorative value of sleep until we have lost it. Getting back into a good sleep pattern is a key stage in the recovery from depression.

Other mood disorders affect sleep in different ways. People with bipolar affective disorder (manic depression) who have an elated mood as part of an episode of mania commonly experience a diminished need for sleep. Other symptoms of elation include enhanced energy, increased pressure and flow of speech, distractibility, disinhibition, grandiosity and a tendency to embark on impractical or extravagant projects or schemes, sometimes resulting in inappropriate romantic or sexual activities, hostile behaviours or reckless decision-making. Ultimately, the lack of sleep becomes problematic in elation and can lead to severe exhaustion. As is the case in depression, sleep problems associated with elation are generally linked to the abnormal mood and tend to improve as the person responds to treatment.

Lack of sleep can also contribute to the risk of developing mental illness in the first place. In the United Kingdom, the National Health Service (NHS) advises that chronic sleep debt may increase risk of long-term mood disorders like depression and anxiety.[10] When people with anxiety or depression were surveyed to calculate their sleeping habits, it was found that most slept for less than six hours per night. The NHS also points out that sleep boosts immunity, helps prevent diabetes, wards off heart disease and increases sex drive and fertility.

It is clear, then, that sleep makes a vital contribution to physical health, mental well-being and happiness in complex, powerful ways. Dean Burnett makes this point in *Happy Brain: Where Happiness Comes From, and Why.*[11] Burnett is a neuroscientist and research associate at the Centre for Medical Education at Cardiff University. He writes about the importance of our homes in shaping our well-being and argues that by allowing us to get enough sleep, our homes increase the chances of us being happy. Burnett cites research showing that we have lower-quality sleep on our first night in unfamiliar surroundings, as we miss the reassurances of home and our brains remain more alert through the night. We sleep best when we are surrounded by the familiar sights, sounds and smells of home.

I can identify with this. Sleep is often more challenging when I am away, at least initially. Travel is tiring but unfamiliar surroundings make sleep more difficult and – strangely – different. I dream more when I am away and that seems to help me sleep, even in unfamiliar surroundings. It usually takes me a night or two to adjust to a new location and to reap the benefits of a good night's sleep once again.

Sonja Lyubomirsky emphasises the links between sleep and well-being in *The How of Happiness: A Practical Guide to Getting the Life You Want.*[12] Lyubomirsky is a professor of psychology at the University of California, Riverside, and a world expert on happiness. She notes that, regardless of how successful, vigorous and active we are in our daily lives, a lack of sleep will have a negative impact on our moods, alertness, energy, health and longevity. Even one more hour of sleep per night can make a significant difference to our health and happiness. I agree: waking at 7 a.m. rather than 6 a.m. makes all the difference to me.

It is clear that there is a strong link between sleep and happiness. The great thinkers of history support this view. Hippocrates noted that 'in whatever disease sleep is laborious, it is a deadly symptom; but if sleep does good, it is not deadly'. Heraclitus recognised that sleep is an active state, with the mind working busily as we slumber: 'Even a soul submerged in sleep is hard at work and helps make something of the world.' And Mahatma Gandhi understood that peaceful, restful sleep is the key to restoring body and mind: 'Man should forget his anger before he lies down to sleep.'

Adding further to this argument, Anthony Seldon highlights the example of cats in *Beyond Happiness: How To Find Lasting Meaning and Joy in All That You Have.*[13] Seldon notes the clear importance of sleep for well-being and points out that cats do not seem to suffer from insomnia, so we should not either. I agree. Certainly Trixie is capable of deep, deep sleep at any given moment and she is clearly very, very happy. Sleep is, perhaps, the secret to her well-being – hours and hours and hours of sleep. For this and many other reasons, we should all be more like cats.

So having established that sleep is good for mind and body, that sleep is associated with happiness, and that we all need to get more of it, what next? What should we actually *do* to better nourish 'life's feast' with more and better sleep?

## How to sleep

Before we discuss practical strategies about how to sleep, it is useful to bear in mind the overarching principles of a happy life that we discussed earlier, especially gratitude, acceptance and balance. When we sleep well we should make a point of being grateful: a good night's sleep is an invaluable gift. When we cannot sleep, we need to accept this fact without anger and try to remedy it calmly: a gentle, considered approach is infinitely better than the frustration that can so often overcome us in the wakeful early hours. And we also need to reach for balance, especially a good balance between sleeping and waking in our lives: both are vital and each nourishes the other.

There are many steps we can take to improve our sleep: the Sleep Council in England and other bodies offer plenty of practical advice,[14] although often these recommendations seem unrealistic and it can feel impossible to adhere to them fully. The best approach is to do as much as you can in the general directions outlined and figure out what works best for you. Everyone is different, so different steps will work for different people. There are, however, some general principles that all of us could usefully bear in mind as we try to improve our sleep.[15]

First, it is important that our bedrooms and sleeping areas are dark and cool. We should keep the temperature around 16–18°C. Our bedrooms should be quiet, uncluttered and free of gadgets. This means removing televisions, computers, tablets, phones and other pieces of electronic equipment. Many people struggle with this as they are accustomed to scrolling through social media in bed at night or using their phones as alarm clocks. All of this is unwise and should be minimised or eliminated. It is perfectly possible to leave your phone in another room or in the boot of the car overnight, so that you are entirely detached from it as you go to sleep. Try this and see how it works. The effects can be dramatic.

It is important that our beds are big and comfortable, and that we change the mattress every eight years or so. Many people do not have complete control over the size of their bed because their budget or the dimensions of their bedroom places limits on them. Even so, it is still important that our beds are as big as we can manage, to give us as much room as possible to move around as we sleep without worrying about falling out of bed or encountering obstacles that might wake us.

Second, lifestyle matters hugely to the quality of our sleep. What we do during the day, especially in the evening, has an enormous effect on how we sleep at night. As bedtime approaches, we should reduce the intensity of light in our homes, using dimmer switches or low wattage bulbs. This helps our bodies make a more realistic, gradual adjustment from daytime to night-time. It is also helpful to establish a pre-bed routine, a fixed bedtime and a regular sleep pattern. A good idea is to set a gentle alarm on your phone for an hour or so before bedtime to remind you to put away the phone and start to wind down.

Other lifestyle tips for better sleep include avoiding alcohol and naps during the day, both of which diminish the quality and quantity of sleep at night. We should also minimise or eliminate the use of computers, mobile phones and televisions in the hours before bedtime. This is a real challenge for people who are accustomed to watching television at the end of a long working day or just before bed. While television can have the benefit of relaxing our minds, it is not helpful for sleep. It is much better to read a (printed) book or to listen to the radio or a podcast in order to wind down.

Third, stress and worry affect sleep. It is important that we both reduce the long-term stresses in our lives and develop short-term strategies to manage our anxieties immediately prior to sleep. Relaxation exercises can help set the scene for a good night's sleep. One useful exercise is the 'body scan', which involves focusing your attention at the top of your head and then moving your focus down your body, systematically noting each part of your body as you progress, and consciously relaxing as you go along. It is important to be systematic, mindful and slow.

If you remain anxious in bed, with your thoughts racing at high speed, one way to calm your worries is by counting sheep or reciting the alphabet as you try to nod off. You can do this in a whisper or, if you have a bed partner, in your head. Repetitive cognitions like these can help us relax and induce a state of pre-somnolence that eases us gently into restful sleep.

Breathing exercises are also helpful: count your in-breaths for ten breaths, so that you have counted from one to ten over the course of ten in-breaths; then count ten out-breaths; count ten turnings of the breath (after the in-breath and before the out-breath); and then start again. This is a common meditation exercise known as 'the mindfulness of breathing'. Practised at night-time, this exercise can help us to relax by redirecting our thoughts away from the events of the day and creating a soothing rhythm in our minds and bodies.

Fourth, diet matters hugely to sleep. A balanced diet, rich in vegetables and fruit, helps to maintain general physical and mental health. These, in turn, promote restful sleep. It is also important to become aware of the effects of specific foods and other substances in the hours prior to bed. Caffeine, alcohol, sugar and cigarettes all hinder sleep. We should minimise these at all times, but especially in the hours before bed. As a rule of thumb, try to avoid stimulants in the evenings. Coffee, in particular, should be limited to mornings, as it can have a long-lasting stimulant effect.

The quantity of the food we eat also matters to our sleep. It is important not to be too full or too hungry when going to bed. The Taoist principle of moderation is important throughout our diet not only because excessive eating is generally unhealthy, but also because overeating contributes to difficulties with sleep. Specific foods affect people differently, so if you think a particular food impacts on your sleep, keep a food-and-sleep diary to identify patterns. In general, foods that are rich in tryptophan often assist with sleep. These include chicken, turkey, milk, dairy, nuts and seeds. Many people drink hot milk before bed, but the usefulness of this varies from person to person.

Fifth, having a good exercise pattern helps with virtually every aspect of life, including sleep. From a sleep perspective, the best time to exercise is in the morning, ideally outside. We will discuss exercise later but, to summarise, adults should get either 150 minutes of moderate aerobic activity (e.g. brisk walking) or 75 minutes of vigorous aerobic activity (e.g. running) each week. We should also do strength exercises on two or more days, working all of our major muscles, including our legs, hips, back, abdomen, chest, shoulders and arms. Exercising in the early morning sends a clear message to our bodies that the day has now started and this will help us to feel tired when it is time for bed in the evening.

Finally, if you go to bed and simply cannot get to sleep, there is a 20-minute rule that can help. If you're lying in bed without sleeping for 20 minutes, get up and read a book for 20 minutes – not an electronic book, but a book printed on paper in the traditional fashion. Do not watch television, drink tea or coffee, smoke cigarettes or look at a phone, tablet or computer. After 20 minutes, go back to bed and try again to sleep. If that does not work after 20 minutes, repeat the procedure until you fall asleep.

It is possible that you will try all of these techniques and still have difficulty sleeping. In this case, if you have a bed partner, it is important to discuss your sleep problems with them. You might need to remove yourself from a shared bed or send your partner to the spare room until your sleep pattern is re-established. Separate beds might also help, as might medical advice about snoring.

If you still have persistent problems sleeping, it is wise to visit your doctor, to see if other issues are affecting your sleep (e.g. hormonal changes or obstructive sleep apnoea). Sleeping tablets or other medications can help, but these need to be taken under medical guidance and in the context of an overall approach to sleep that includes good sleep hygiene.[16] Sleeping tablets are never the answer on their own and ideally should be used only for short periods when necessary.

Overall, improving sleep involves a range of changes to our habitual behaviours both during the day and at night-time. The key lies in convincing our bodies that bedtime is the time for sleep and morning is the time for getting up and getting on with the day. It is this balance between sleeping and waking that determines how refreshing we find our sleep at night and how satisfying we find our days.

Ultimately, we need to focus on both falling asleep and waking up in order to find the optimal balance between sleep and wakefulness. Both sleeping well and waking up matter greatly to our happiness. Having discussed sleep, let's now consider waking up.

## How to wake up

In 1955 Jack Kerouac, the American novelist, wrote a biography of Siddhartha Gautama, better known as the Buddha, the spiritual leader, teacher and foundational figure of Buddhism. The word 'Buddha' actually means 'Enlightened One' or 'Awakened One', so Kerouac called his book *Wake Up: A Life of the Buddha*.[17] Tragically, Kerouac died in 1969 before his biography of the Buddha was published. The book was later serialised in *Tricycle: The Buddhist Review* between 1993 and 1995 and eventually published in book form in 2008. *Wake Up* is a fascinating read, revealing much about Kerouac as well as telling the story of the Buddha with empathy and understanding (just as the Buddha lived his life).

The title *Wake Up* captures one of the central messages of Buddhism: that we should wake up to the true nature of reality around us and within us. We should step away from delusion and try to see things as they really are, without illusion, denial or desire. Only then can we be free of the ties that bind us to endless cycles of attachment, disappointment and disillusionment. Only then will our eyes be opened. Only then will we be free.

Each day represents a new opportunity to achieve these goals. In the words of the Buddha: 'Today's the day to keenly work – who knows, tomorrow may bring death!' Other spiritual traditions and religious texts offer similar advice. In the Bible, the Book of Isaiah advises the reader to 'wake yourself, wake yourself, stand up'. The Gospel of Luke recommends that we 'stay awake at all times'. And Psalm 57 is full of wakefulness: 'Awake, my glory! Awake, O harp and lyre! I will awake the dawn!'

But do all these exhortations to awaken mean waking up in the morning, getting out of bed and brushing our teeth? Or do they mean waking up in a spiritual sense, acknowledging the nature of reality and finding a higher purpose in life? Or might they possibly refer to both kinds of awakening – waking up in the morning *and* awakening in our souls? It is worth thinking about this a little.

I have often wondered why monastic life tends to involve getting up so early. Many Buddhist monks rise at 4 a.m., meditate, chant, gather alms and do all sorts of household duties before they have breakfast. When I wake at 6 a.m., I am invariably ravenous and can do nothing until I have eaten. I race downstairs. There is not the faintest possibility of me exercising, meditating, chanting, gathering alms or achieving any kind of spiritual awakening until I'm up, breakfasted, dressed and organised for the day. Kerouac, despite his spiritual interests and biography of the Buddha, scarcely adhered to monastic discipline either, although his problems were different from mine: he was plagued with restlessness, drugs and heavy drinking, which contributed to his early death at the age of 47.

But even if getting up early was not a steady feature of Kerouac's life, perhaps *waking up* was. Kerouac was wide open to new experiences and always ready to engage in spiritual searching through how he lived his life, the friendships he forged and the writings he produced. Kerouac might have lacked the day-to-day discipline of a Buddhist monk, but discipline is not necessarily an end in itself: awakening or enlightenment is.

In 2014 Sam Harris, neuroscientist and writer, wrote an interesting book that touches on these themes, called *Waking Up: Searching for Spirituality Without Religion*.[18] Harris argues that there is more to understanding reality than science and secular culture usually permit, and that how we pay attention to the present moment largely shapes the quality of our lives. A *New York Times* bestseller, Harris's book combines stories from his life with selected aspects of various spiritual traditions, discussions about neuroscience and reflections on the mystery of consciousness. In the end, Harris recommends that we seek direct experiences of reality and truly wake up to the reality that is all around us and within us.

Harris is, of course, correct in his recommendations, but perhaps the most interesting feature of his advice is how firmly it belongs to a long and noble tradition of ideas along these lines. There are elements of this approach to be found in early Buddhist mindfulness traditions, the nature writings of Henry David Thoreau and the recent avalanche of books about mindfulness. If consistency is what we seek, then the advice that we should 'wake up' rates very highly indeed: it is everywhere.

But how can we 'wake up' more in our lives and worlds? What should we actually *do* in order to awaken?

The first point to note is that the Buddhist monks are right: getting up at a set time every morning, ideally relatively early, sets the scene for better engagement with each day and with the world in general. We do not need to rise at 4 a.m. and meditate for an hour, but getting up reasonably early at a fixed time will help us to develop a more disciplined, focused approach to life. This enhances both our days and our nights, and can be especially helpful for sleep. Buddhist monks usually sleep soundly, I am told.

The second piece of advice is that when we wake up in the morning, we should get up fully and start the day. Earlier in this chapter we discussed the value of having a good exercise pattern, on the basis that moderate exercise helps with every aspect of our lives, both during the day and when we want to rest at night. From the point of view of sleep, the best time to exercise is in the morning, ideally outside. By this logic, once we wake up, we should jump out of bed, rush outside and exercise immediately. In truth, few of us do this, but we can still try to move more in this direction.

How can we do this? The first step is to get out of bed when we wake up. Lying in bed checking our phones in the morning is immensely unhelpful. Our phones should be downstairs or in the boot of the car from the night before: phones should never enter bedrooms. Snooze buttons on alarm clocks are also problematic: they delay our engagement with the day and likely set back the time at which we will fall asleep at bedtime. Once awake, we should get out of bed and get on with the day.

Second, even if we do not race outside to exercise, there are plenty of other things we can do to tell our bodies that the day is now beginning. If we have a garden, we can step outside for a few moments as we have a cup of tea and let the morning light hit our faces. This sends a clear message to our bodies that we are now awake. The earlier in the day that our bodies get this message, the better we will sleep at night-time.

These are many steps we can take to ensure that when we wake up, our bodies are fully clear that the day has commenced. We can be more generally active in the mornings, practise some meditation, walk to work instead of driving, or simply try to spend more time outside. These are not explicitly spiritual steps towards inner awakening and attaining enlightenment, but they are important moves towards embracing reality, just as so many spiritual and philosophical traditions recommend. If we take these small actions, spiritual progress will silently follow. Enlightenment grows.

Renowned teacher and researcher Jon Kabat-Zinn makes a similar point in *Full Catastrophe Living: Using the Wisdom of Your Body and Mind to Face Stress, Pain, and Illness.*[19] Kabat-Zinn points out that if we make a commitment to be fully awake when we are awake, then many of our sleep problems will likely resolve themselves. Some mediation exercises, such as the body scan that we discussed earlier, can, paradoxically, help us to fall asleep at certain times and to wake up at others. The key is that the exercise helps us focus on the present moment and accept whatever is in that moment, be it sleepfulness, wakefulness or whatever. We cannot always change what is in the moment, but we can identify it, experience it and navigate it mindfully, with self-compassion.

Many of the techniques that help us to wake up fully in the morning are also recommended for spiritual awakening: focusing on the moment, embracing reality, deepening awareness and coming closer to nature. The Buddhist monks who rise at 4 a.m. and Sam Harris are correct. We should fall asleep when we are tired, wake up fully in the morning and engage directly with ourselves and the world.

The importance of sleep is often overlooked in our rush to be happier. This is a mistake. Good sleep is fundamental to well-being. Without sleep, there is no lasting happiness.

Just as we need to learn to sleep, we need to learn to wake up properly in the morning so that our bodies are in no doubt that the day has now begun. Early morning exercise or even just stepping outside can be helpful. Waking up also means being alert to each moment throughout the day, committing to being fully awake when we are awake, and fully asleep when we sleep. Sleeping and waking are two sides of the same coin, so we rarely have one without the other.

Achieving a balance between being asleep and being awake is the key. In the words of Leonardo da Vinci, 'a well-spent day brings happy sleep'. This works both ways: just as a well-spent day brings happy sleep, happy sleep brings a better day.

Sometimes, the problems that we link with sleep do not come from failing to sleep; they come from failing to wake up. We need to do both to be happy.

## Top tips for happiness: sleep

- If you are an adult, aim to sleep for between 7 and 9 hours in every 24 hours.
- Exercise during the day, ideally outside and in the morning.
- Enrich your diet with foods that contain tryptophan (e.g. chicken, turkey, milk, dairy, nuts, seeds).
- Avoid naps during the day.

- Avoid stimulants in the hours before sleeping (e.g. coffee, alcohol, sugar, cigarettes).
- Ease yourself into sleep in the evening: dim lights, avoid alcohol and screens, and wind down gently before bed.
- Keep your bedroom dark, cool, comfortable and free of distractions (especially screens).
- In bed, relaxation exercises or breathing techniques can help to settle your racing thoughts as you try to sleep.
- If you cannot sleep, use the 20-minute rule: try to sleep for 20 minutes; if that does not work, get up and read a book for 20 minutes; try again to sleep; and if that does not work after 20 minutes, repeat the procedure until you fall asleep.
- When you wake in the morning, wake up fully and start the day with gratitude for your sleep.

FIVE

# Dream/Stop Dreaming

Most people dream about becoming rich. People who are already rich dream about being richer. And those who are very rich dream about being happy. What is going on?

The happiness research we discussed in Chapter 2 told us three things about income and happiness. First, we all need a basic level of income to support our well-being. If our fundamental human needs are not met, it is highly unlikely that we will be happy. Second, increasing income is associated with greater well-being and happiness, but only to a certain point. Income beyond around $95,000 per year is unlikely to bring us significant additional happiness. Third, it is entirely possible that income above $95,000 will actually decrease rather than increase our well-being. This means that fabulous wealth not only has diminishing returns, but might well make us unhappy in the end.

I am fully aware of all of these facts and yet I frequently dream and daydream about becoming incredibly wealthy. Why? Why are my dreams and daydreams an evidence-free zone, where facts that I understand and accept are tossed aside in favour of irrational fantasies, self-defeating imaginings, bizarre happenings, distorted time frames and goodness knows what else? Do they make me happy or unhappy? If they make me unhappy, why do I still have them? If they make me happy, can I do anything to help them make me happier?

People have always been intrigued by dreams and what, if anything, they tell us about ourselves. Leonardo da Vinci asked: 'Why does the eye see a thing more clearly in dreams than the imagination when awake?' Why do events in dreams somehow feel more real than events in our waking lives?

Henry David Thoreau said that 'dreams are the touchstones of our character', reflecting the deeply held belief that they provide a special window into our true selves (as if our waking selves are somehow less than 'true'). Ralph Waldo Emerson agreed: 'Judge of your natural character by what you do in your dreams.' If we are to judge my 'natural character' on the evidence of my dreams, then I am extremely odd, highly illogical and deeply unpredictable! I believe I am not alone in this.

The possible links between dreams and reality have intrigued the great philosophers of history. Aristotle said that 'hope is a waking dream'. Plato wondered if we can ever know when we are dreaming and when we are awake: 'How can you prove whether at this moment we are sleeping, and all our thoughts are a dream; or whether we are awake, and talking to one another in the waking state?' Johann Wolfgang von Goethe advised us to 'dream no small dreams for they have no power to move the hearts of men'.

We know that people who are prevented from dreaming at night develop psychological problems quite quickly; and people who dream too little or too much during the day lose touch with reality in a different way. So how can we achieve a balance between these two positions and increase our happiness through our dreams and daydreams?

## What happens when we dream?

Everyone has a different theory about where dreams come from. Philosophers explain dreams using the tools and concepts of philosophy, rooted in their particular training and view of reality. Psychologists provide psychological theories and cognitive explanations that reflect the way they choose to look at the world. Artists appeal to the imagination, seeing multiple visions and dimensions in dreams, often stretching into realms far distant from those ever envisaged by the dreamer. And, in recent years, scientists have leapt into this rather crowded field, seeking a better biological understanding of what is happening in our brains when we dream. Let's start with the science.

There has been a significant increase in scientific research about dreams in recent decades, starting with the observation that certain types of brain surgery that are now rare, such as prefrontal leucotomy, often impaired the ability to dream, confirming a biological basis to dreams.[1] Later research sought to link dreams with specific anatomical areas in the brain and, while this approach was interesting, it has been largely supplanted by more recent work that looks at brain systems or networks involving multiple areas across the brain, rather than searching for a single 'dream centre'. This shift has seen the field move forward significantly.

This more recent approach also fits well with what is known about how the brain works in general and the physiology of sleep and dreams in particular. Humans experience two types of sleep: rapid eye movement (REM) sleep and non-REM sleep. Each type of sleep is linked to specific brain waves and patterns of brain activity. We experience both types of sleep during any given night, with more REM sleep occurring as the night progresses. We dream in both REM and non-REM sleep, but dreaming is more common in REM sleep. REM dreams tend to be more vivid and bizarre, and if we wake from an especially vivid dream, it is likely that we have woken from REM sleep. Non-REM sleep is associated with more thoughtful, slower dreams.

In terms of biology, REM sleep is associated with activation of specific brain areas such as the limbic system (which is concerned with emotion) and deactivation of other parts of the brain such as the prefrontal cortex, which is an area near the front of the brain that controls most of the mind's internal imagery when we are awake. These observations confirm what many of us know from our own dreams: emotions can run high in dreams and imagery can run wild. Summarising the evidence as a whole, it is now clear that dreaming is a state of consciousness that is associated with reduced controls and constraints on memory and perceptual imagery. Thus, our dreams can be vivid and devoid of the usual regulatory controls that keep our waking thoughts within certain parameters. When we dream, we are free.

Matthew Walker discusses the biology of dreams in some detail in *Why We Sleep: The New Science of Sleep and Dreams*.[2] He notes that recent brain imaging research links REM dreams with activation of brain regions associated with complex visual perception, movement, autobiographical memory and emotion. Areas concerned with emotion can be up to 30 per cent more active during REM sleep than when we are awake. In addition, Walker confirms that some of the prefrontal areas of our brains are less active in REM sleep, thus switching off the regulatory function that keeps our daytime thoughts rational and organised. REM dreams see our emotional and visual brains freed from the constraints of our waking hours. In dreams, anything and everything can happen (and frequently does).

Owing to the marked differences in brain activity between waking and sleeping, it is scarcely surprising that we can have unusual experiences during transitions from waking to sleeping and sleeping to waking. These experiences are common and were described by such early thinkers as Aristotle. Many of us are familiar with a sudden jerk in one of our limbs as we fall asleep or suddenly becoming fully awake again with an abrupt feeling that we have just fallen. These experiences are known as hypnagogic phenomena, which means that they occur during the period of drowsiness that immediately precedes sleep, as we are nodding off.[3]

Hypnagogic hallucinations are three times more common than hypnopompic hallucinations, which are similar experiences that occur just as we are waking. For example, we might 'hear' the sound of a phone ringing and believe that this has woken us, when, in fact, the ringing was a hallucination that occurred as our brains transitioned from sleeping to waking. It is important to recognise that hypnagogic and hypnopompic phenomena are generally nothing to worry about, although they can occur in narcolepsy, which is a sleep disorder characterised by excessive sleepiness, hallucinations, sleep paralysis and, for some people, episodes of cataplexy (loss of muscle control). Mostly, however, hypnagogic and hypnopompic experiences just feel a bit odd and have no medical significance.

It is interesting to compare hypnagogic and hypnopompic phenomena with dreams. One of the most intriguing aspects of hypnagogic and hypnopompic experiences, such as hearing a voice while you fall asleep or wake up, is that such perceptions do not form part of a continuous experience in which we feel we participate. They seem to occur suddenly and come from outside of us. This differentiates these experiences from dreams, in which we are clearly participants, observers or (weirdly) both. Dreams have a curious emotional intensity in which we are usually deeply invested even if the contents of the dream itself do not, on the face of them, seem to merit sustained emotional involvement. No matter how irrational they are, dreams matter deeply as we have them.

This intensity remains one of the many mysteries of dreams. It is unfortunate that our biological understanding of dreams is sharply limited by our rudimentary knowledge of the brain itself – science still struggles to describe how the brain works when we are awake, let alone when we sleep and dream. The science of dreams, such as we know it, does make sense, insofar as it goes, but it does not come close to explaining why we dream in the first place. Why does all of this happen while we sleep? It all seems complex and highly unlikely, if I am honest.

We know that dreams are an integral part of sleep and that sleep is necessary for both physical and mental health. But why is something as peculiar as dreaming necessary in order for sleep to deliver these benefits? As I reflect on this question Trixie lies fast asleep on my writing desk, her ears and nose twitching from time to time. Do cats and other animals dream and, if they do, what do they dream about? If they don't dream, is dreaming part of what sets humans apart from other animals? And can we ever really know the answers to these questions?

The first question is whether or not all mammals have dreams the way that humans do.[4] It turns out that, if we take a hard stance on the issue, it appears unlikely that certain creatures such as monotremes (e.g. platypuses) or cetaceans (whales, dolphins and porpoises) experience REM dreams. Atypical REM sleep in certain other species, such as Arabian oryx and African elephants, might alter their potential to experience REM dreams. The fact that humans dream in both REM and non-REM sleep, however, means that all mammals might well have the potential to dream, albeit that non-REM dreams might be less complex in certain mammals compared to others. Overall, then, it is entirely possible that all mammals, and maybe even all animals, dream.

Even bees, it seems, experience different stages of sleep and might well have dream-like experiences that are similar to non-REM dreams in humans.[5] If this is the case, it is possible that all sentient beings have the potential to dream. Indeed, humans often blithely assume that our dreams and imaginations are especially complex compared to those of other animals, but this might not be so. I have no idea what Trixie dreams about, presuming that she dreams at all – she might have magnificently elaborate dreams, vastly surpassing my own.

Overall, it is clear that dreams are quite mysterious, in both humans and other animals, but their ubiquity and intensity suggest that they might be linked in some way with sentience or consciousness. Or, to put it the other way, an absence of consciousness likely indicates an absence of dreams. To return to Aristotle: 'Nothing is what rocks dream about.' Humans dream because our brains are big and complex, and dreaming is somehow intrinsic to sleep.

So I think I can conclude that Trixie is indeed dreaming cat dreams as she sleeps here beside me on my desk. The fact that she cannot tell me about her dreams makes it difficult for me to be definitive on this point, but research certainly suggests that Trixie has the potential to dream. If this is the case, she probably dreams quite a lot, given the amount of time she spends asleep. But do Trixie's dreams make her happy, over and above the role of dreams in maintaining sleep itself, which is essential for well-being in its own right? Do dreams add additional happiness to sleep, for either Trixie or me?

In terms of the science, it is clear that human dreams, especially vivid REM dreams, involve the activation of brain regions associated with complex visual perception, movement, autobiographical memory and emotion, and the deactivation of the regulatory function that usually keeps us rooted in reality. As a result, do dreams offer a kind of imaginary freedom from restraint that is necessary for happiness? And, if they do, can we do anything to optimise this and increase the happiness that we glean from our dreams?

## What do our dreams mean?

Paracelsus, a 16th-century Swiss physician, wrote that 'the interpretation of dreams is a great art'. If Paracelsus was right about this, then the Austrian neurologist Sigmund Freud was one of the most imaginative artists of all time. Born in 1856, Freud is hailed as the father of psychoanalysis and one of the foremost thinkers of the twentieth century. Freud was deeply interested in dreams and how we can use them to better understand ourselves and resolve the various problems in our lives. According to Freud, dreams provide a unique insight into the human unconscious, a window to our souls.

Freud's magnum opus on this theme, *The Interpretation of Dreams*, was published in 1900.[6] In the book, Freud argues that dreams are not meaningless or absurd, but rather psychological experiences of complete validity. Moreover, he says that dreams represent the fulfilment of wishes: desires that are not realised in our waking lives can be fulfilled in dreams, even though this is not always immediately apparent when we think back on our dreams.

Freud's theory makes obvious sense for pleasant dreams in which we see our wishes clearly fulfilled, but it does not readily explain why so many dreams are unpleasant, nonsensical and generally unsatisfactory. Freud has an answer for this, too, saying that we must differentiate between the *manifest* and *latent* content of dreams. Interpreting dreams requires our focused attention because dreams *are not made with the intention of being understood*. Dreams come from a subconscious that is indifferent to whether or not our conscious minds understand what is happening. Therefore, to understand what our dreams mean, Freud advises that we need to dig beneath the surface and actively interpret their content.

The initial reception of *The Interpretation of Dreams* was lukewarm at best. Only 351 copies were sold in the six years following publication. Over the following decades, however, the book gradually became one of the classic texts in psychology and psychoanalysis. There are many possible reasons for this, including the systematic approach that Freud brings to the topic of dreams, the enthusiasm with which he presents his theories and the compelling style with which he conveys his thoughts. But most of the book's popularity rests on our own innate belief that our dreams have important hidden meanings, if only we could figure them out. Is this true?

Freud's idea that dreams represent wishes being fulfilled is a helpful place to start, even if this is unlikely to be a universal rule. Freud is, perhaps, on firmer ground when he simply asserts that dreams are psychological experiences of substantial validity and their content relates to our waking lives in interesting ways that are not always immediately apparent. But, if this is the case, how do our sleeping brains select which material to put into dreams and which material to omit? And what is the meaning in this selection process?

To investigate this, one research group looked at the incorporation of recent waking-life experiences into dreams in an effort to figure out which material our brains select for inclusion.[7] To do this, researchers woke study volunteers from REM and non-REM sleep and obtained dream reports from them. The content of the volunteers' dreams was compared with log records of their previous daily experiences to see which waking-life experiences made it into their dreams and which did not. The researchers also monitored brain activity using electro-encephalography (EEG), which records electrical activity in the brain.

This study found that the number of references to recent waking-life experiences that featured in REM dreams was linked with frontal brain activity during REM sleep. There was no such association for older memories or for non-REM dreams. In addition, the emotional intensity of recent waking-life experiences incorporated into dreams was higher than the emotional intensity of experiences that were not so incorporated. These results indicate that dreaming reflects emotional memory processing during REM sleep and that the most emotionally charged memories of recent days are prioritised in this process.

So what are the implications of these findings? First, these results support the view of Freud and many others that the selection of material for inclusion in our vivid REM dreams is not random: while the content or storyline of any given dream can appear haphazard and arbitrary, it is not. There is a logic at work that informs our dreams, includes certain material and omits other material. Dreams are not just accidental by-products of our over-busy, over-complicated minds. Dreams have significance. We might not immediately see what that significance is, but our dreams do mean something.

Second, these findings provide evidence that vivid dreams prioritise emotionally charged material in need of further processing and deprioritise less emotionally charged material – this suggests that dreaming acts as a form of overnight therapy that smooths out the emotional roller-coaster of our waking lives into a more acceptable and meaningful format.[8] As we sleep, we work through emotions in our minds and, in doing so, generate dreams that are sometimes bizarre, frequently intense and generally helpful for our well-being.

Having established that dreams are not random events and that recent, emotionally charged material is prioritised for inclusion, can we figure out what our specific dreams mean and use that knowledge to increase our happiness?

There are several helpful messages that we can take from the literature about dreams and apply in our own lives. The first is a simple one: we should not ignore our dreams. Our dreams have significance and meaning and, as a result, they merit our attention. What we do with this information is up to us, but, broadly, if we ignore our dreams, we miss an important opportunity for self-understanding. We should listen to our dreams, even if we do not understand what they are saying to us.

Second, the meaning of certain dreams is very obvious. Some people 'dream' of a telephone ringing and wake to find that a telephone really is ringing, and that the sound of the telephone was incorporated into their dream just before it woke them. Such dreams are straightforward and do not require deep introspection.

Other dreams concern events that really occurred: positive or negative childhood experiences, recent triumphs or disappointments, or the fulfilment of known wishes (such as me dreaming about being fabulously wealthy). It is still useful to reflect on these dreams after we have them. The fact that they are readily understandable does not make them less important. Often our minds focus on our mysterious, inexplicable dreams rather than acknowledging the clear, simple messages from our more explicable ones. If you routinely dream about how much you dislike your job, there is a message there. We need to listen to our obvious dreams, as well as our more mysterious ones.

The third useful message from the literature about dreams concerns our more obscure dreams and how we might figure out what they mean. Freud's distinction between the latent and manifest content of dreams is most relevant to dreams that trouble us and seem, on the face of them, to make no sense at all. While these dreams are certainly worth thinking about, it is vital that we do not take every detail literally. Obscure dreams are presented in densely coded forms that require interpretation. Some people tie themselves up in knots trying to interpret every detail of such dreams literally.

When our brains create dreams, they condense material, move things around, telescope time frames and shift perspectives. As a result, dream language is different to waking language, so we should focus on the broad, recurring themes in dreams rather than minutiae. If you dream about a friend giving you some money, the themes here are friendship and wealth. The details of the transaction are not necessarily important. Look for the big themes in dreams and interpret them broadly.

It is good to be aware of specific types of dream that can occur and some of the possible contributing causes. Stress, trauma, medications, drugs, alcohol and migraine can all be associated with nightmares.[9] Night terrors, which are intense episodes of fright in the midst of dreams that occur in up to 6 per cent of children, can be linked with sleep apnoea and enlarged tonsils and adenoids, among other conditions. These require medical assessment. There are also 'lucid dreams', in which the dreamer is reportedly aware that they are asleep and can sometimes shape their dream as it unfolds. This is rare.

All told, the key message from research about dreams is that our dreams are not random events but part of the way that our minds process emotional material. This is important for our happiness and well-being, so we should listen to our dreams, even if we do not fully understand what they are saying. Be wary of grand interpretative theories. While many dreams clearly reflect the fulfilment of wishes, as Freud suggested, other dreams likely represent other things. Some meanings can be very simple. Accept them.

Finally, we should reflect on the key themes in our more mysterious dreams, but skip over many of the details, which can simply be part of how dreams are stitched into narratives rather than part of their core purpose. Not everything is laden with meaning. The principle of acceptance is important here. As Freud is reported to have said (but probably did not): 'Sometimes a cigar is just a cigar.' So too is it with dreams. But what about daydreams?

## Daydreams

What about those times during the day when we purposely let our minds wander and spend a few minutes imagining things that we know are not real? Does this increase or decrease our happiness? I can see arguments both ways.

One research group analysed data from 2,250 adults in the United States, asking them if they were, at that moment, thinking about something other than what they were doing.[10] The question was asked at random times during waking hours using mobile-phone technology. The results showed mind-wandering in some 47 per cent of responses, suggesting that we spend almost half of our time thinking about things other than what we are doing. As often as not, it seems, our minds are entirely elsewhere.

Of course, not all mind-wandering is true daydreaming, but the researchers found that people's minds are more likely to wander to pleasant topics rather than unpleasant or neutral ones. Much mind-wandering is therefore likely to be daydreaming, even if we are only fantasising about what we will do when a particularly tedious meeting finally ends. Interestingly, these researchers also asked study participants about their happiness levels and found that people are happier when their minds are *not* wandering and when they are focused on what they are doing. In other words, when we are unhappy with what we are doing, our minds drift off.

This research confirms that our minds wander a lot, often to positive topics. The fact that we are less happy when our minds wander does not mean that a wandering mind *makes us* unhappy. It is far more likely that our mind wanders *because* we are unhappy in our present activity and we seek to escape, if only in a daydream. This makes sense: my mind wanders continually when I am bored or detached from what I am doing. I sometimes deliberately let this happen and it helps me get through some uninteresting meetings. I enjoy daydreams and have several favourite ones that I routinely indulge in. I do this because I think that the benefits of daydreaming exceed the costs. But do they? Should I worry about my drifting mind?

Perhaps. There is evidence that mind-wandering occurs at a significant cost to our performance on tasks such as reading, sustained attention and tests of aptitude.[11] Mind-wandering can also have a negative impact on reading comprehension and model-building, reduce our ability to withhold automatised responses in tests and disrupt our performance on assessments of working memory and intelligence. All of these negative consequences must, however, be weighed against recent evidence that mind-wandering can also have significant benefits. Chiefly, there is now good reason to believe that mind-wandering plays an important role in autobiographical planning and creative problem-solving, both of which are essential psychological and emotional processes in our inner lives. Both of these activities, it seems, might benefit from some judicious daydreaming.

These findings make sense and fit with many people's experience of mind-wandering in general and daydreaming in particular. Most of us are fully aware that we concentrate less when we are daydreaming, but this is often a price that we are willing to pay. If I am in a situation where I don't need to concentrate at all, I am happy to let my attention drift. In these circumstances, I don't mind paying the cost of poor attention and poor retention if I get the pleasure and benefit of daydreaming. I quite enjoy the mildly subversive thrill of focusing my attention elsewhere as my thoughts drift away.

We also know that the wandering mind is generally future-oriented. We plan our lives in daydreams. Admittedly, the future that we daydream for ourselves might be optimistic to the point of fantastical, but at least it is a forward-looking plan. Even better, by releasing us from the pesky constraints of reality, daydreams allow us to think more creatively about the problems of today and the possibilities of tomorrow. This facilitates greater imagination, better problem-solving and reaching conclusions that our rational minds would never permit if we were concentrating fully. Like dreams at night-time, daydreams set us free.

But what is happening in our brains when we daydream? Are our brains awake, asleep, or a mixture of both? Or are they in some entirely different state that is specific to daydreams?

To figure this out, one research study looked at the brain scans of 15 volunteers as their minds wandered.[12] The findings showed that, contrary to expectations, numerous brain regions become active when our minds wander. In fact, our brains are more active when our minds wander than when we are focused on routine tasks.

Up to this point, it was thought that, when our minds wandered, the only part of the brain that was active was its 'default network', which is associated with low-level, routine mental activity. This network is linked with specific brain areas such as the medial prefrontal cortex, the posterior cingulate cortex and the temporoparietal junction. Study results, however, revealed that the brain's 'executive network' is also activated as we daydream, involving other brain areas such as the lateral prefrontal cortex and the dorsal anterior cingulate cortex. This network is concerned with complex, high-level problem-solving.

This means that our routine brain networks and our 'executive' networks are both busy when we daydream, making our minds paradoxically hyperactive as they drift away from our chosen task. As a result, our minds are far from idle when we daydream. Quite the opposite: they are fully occupied with important, complicated tasks.

This finding greatly increases the value that we should place on daydreaming. We might not fully know what our brains are doing as our minds wander, but we can be assured that they are working away furiously. We might lose focus on the task that we set out to do, but that is simply our brain's way of telling us that it has more important things to think about: our relationships, our future goals or maybe just some general reflection. Whatever it is, we should let our minds wander off every so often and see where they take us. For some people, doing an activity helps with this: walking, running, swimming. For others, just sitting quietly allows their thoughts to drift away more easily. Our minds know what they are doing. We should trust them more.

And what about our daydreams? Should we take them more or less seriously, or more or less literally, than we take our dreams? We are awake when we daydream, so does that mean that our daydreams are more meaningful than our dreams or less meaningful?

From a psychoanalytic perspective, daydreams can act as defence mechanisms and can also share the wish-fulfilment aspect of dreaming, even though daydreams occur when we are awake.[13] Analysing the content of daydreams can be helpful in psychotherapy, not only by deepening the therapeutic alliance but also by providing a framework for analysis. This view is highly consistent with the brain scan research demonstrating that daydreaming is a complex brain activity of considerable validity, not unlike dreams themselves. And, like dreams, daydreams invariably mean something, even if we don't quite know what.

The source, form and function of daydreams can all be usefully considered during the course of therapy. While some daydreams present a defence against certain aspects of reality, they can also represent the fulfilment of wishes and so provide an important window into our inner lives – just like dreams themselves. As a result, we should let ourselves daydream at certain times. We should also reflect on the content of our daydreams and, while we should not take them literally, we should be mindful of repeated key themes. Daydreams occur for a reason.

Daydreaming is both valuable and odd, not least because it is a form of dreaming that we consciously allow to happen while we are awake. Some of us (like me) actively encourage daydreaming. We know that our minds are drifting and we deliberately enter a fantasy world for a period of time. We entertain unreality in order to make reality more bearable and – it turns out – to help us work on problems or activities other than the ones we are supposed to focus on. Daydreams are essentially our brain's way of telling us that we have more important things to do. We should listen to this message.

## Reaching a balance

I grew up in Galway in the west of Ireland. For as long as I can remember, we had a poster on the kitchen door with a favourite quote from Albert Einstein: 'Imagination is more important than knowledge.' Many years later, when I got my own home, I found a poster with the same words and today it occupies pride of place on the stairs in my house. The same image also graces my office wall and, some years ago, I dug out the full text of Einstein's quote: 'Imagination is more important than knowledge. For knowledge is limited to all we now know and understand, while imagination embraces the entire world, and all there ever will be to know and understand.'

Imagination is a key element of human thought and a vital part of what it means to be human. When we are awake, we imagine things in many ways, both when we consciously look for new solutions to old problems and when we daydream about how we would like the world to be, rather than the world as it is. Our dreams at night are generally highly imaginative too, sometimes bizarrely so, as the reality-testing of daytime is suspended and anything can happen in dreams (and often does).

We cannot, however, spend our entire lives in a dream, so we need to achieve a balance between dreaming and remaining focused on reality. Balance is one of the key principles of a happy life. This is not too difficult to achieve, but it does require some conscious thought.

The first step is to ensure that we get enough sleep in the first place. Dreaming too little is probably worse than dreaming too much. While sleep is vital for many aspects of our mental and physical well-being, so is dreaming, and we should make certain that we get enough high-quality sleep every night to ensure that we have plenty of opportunity to dream. Most adults do not achieve the seven to nine hours sleep recommended by the National Sleep Foundation.[14] This is a pity. If we don't sleep enough, we can't dream enough, and we miss out on many of the mysterious benefits of dreams.

Second, we should pay attention to our dreams, but not obsess about them. Again, balance is key. Dreams that we perceive as nonsensical can harbour important messages, in the same way that much of what we perceive as reality can prove less sensible and reliable than it first appears. Edgar Allan Poe wrote that 'all that we see or seem is but a dream within a dream'. The demarcation between dreams and reality is not always as crisp as we like to think: we should ignore neither dreams nor reality, in order to get the fullest picture possible about what's really going on. Keep paper and pencil by your bed to note your dreams as soon as you wake up. Dreams can be usefully seen as gifts from our unconscious that we should not ignore.[15] Vincent van Gogh understood this: 'I dream of painting and then I paint my dream.'

Third, we should be aware of both the benefits and the costs of daydreaming. It is useful to be clear at any given time whether we want to focus on what we are doing or whether we are happy to let our minds drift away. Both pursuits are entirely reasonable, depending on the circumstances. It is helpful to consider the idea of 'mindful daydreaming', consciously deciding to divert our attention from what we are doing in order to let our minds wander. This is an excellent way to get through tedious, protracted situations of various sorts. Daydreaming also facilitates imaginative and indirect problem-solving as our minds head off towards whatever topic they feel most needs their attention. This is not appropriate at certain times, such as when we are driving, but it is highly recommended at others.

Fourth, we should reflect on the content of daydreams just as we reflect on the content of night dreams. Daydreams are far from random and our brains are busy as our minds wander. To return to Edgar Allan Poe: 'Those who dream by day are cognisant of many things which escape those who dream only by night.' Daydreams offer boundless possibilities for problem-solving and future-planning, and also perform various other enigmatic functions that our conscious brains struggle to understand or even describe. We ignore daydreams at our peril.

Of course, waking from our dreams is just as important as dreaming in the first place. When we wake, we should awaken fully. When we return our attention to the tedious meeting, we should concentrate fully on that, set our daydreams to one side and possibly think about them a little more later on. The trick lies in achieving a dynamic balance between dreaming and waking, and maintaining an awareness of when we decide to drift off into dreamland.

Waking up more broadly to the nature of reality is also important. Jack Kerouac emphasised this in the title of his book about the Buddha, *Wake Up*,[16] and Sam Harris highlights it in the name of his book about searching for spirituality without religion, *Waking Up*.[17] This idea of wakefulness, alertness, engagement and awareness is a valuable counterpoint to our dreaming and daydreaming. Both dreaming and waking states matter deeply to our happiness, so we should cultivate both with gusto.

Recent years have seen a regrettable trend towards a tendency to dismiss our dreams. In the third of his *Five Lectures on Psycho-Analysis* presented at Clark University in Worcester, Massachusetts, in September 1909, Freud pointed out that people in the ancient world did not ignore their dreams or dismiss them as incomprehensible,[18] and neither should we. There is value embedded in our dreams and daydreams, even if we struggle to recognise precisely what they mean at the time.

Part of the problem lies in the perplexing nature of dreams, which are often full of unusual perspectives, altered time frames and pure imagination. But, as Einstein pointed out, 'imagination embraces the entire world, and all there ever will be to know and understand'. That would be a lot to miss out on by simply ignoring our dreams. The principles of acceptance, gratitude, balance and belief are all relevant to dreams.

We do not dream enough and we do not sufficiently value our dreams. We need to fix this if we are to realise the potential of dreams to chart a happier future for ourselves. To return to Goethe: 'Whatever you can do, or dream you can, begin it. Boldness has genius, power and magic in it.'

## Top tips for happiness: dreams and daydreams

- Sleep, dreams and daydreams are all important for happiness.
- Dreams are full of meaning, even if the meanings are largely obscure at the time.
- Reflect on key themes that recur in dreams, but do not obsess over every detail.
- A dream diary can help; note your dreams after you wake up.
- Accept the intrinsic mysteriousness of dreams: our inner minds can be very odd indeed.
- The more we accept the strangeness of dreams, the happier we will be.
- Daydreams matter just as much as night dreams.
- Our brains are active when we daydream, engaged in complex thought.
- Let your mind wander at appropriate times: daydreaming during a tedious meeting is perfectly acceptable and even recommended; daydreaming while driving a car is not.
- Balance is essential: there are times to dream and daydream, and times to focus on reality.

SIX

# Eat/Stop Eating

Food matters. Napoleon famously said that 'an army marches on its stomach'. George Bernard Shaw held that 'there is no love sincerer than the love of food'. Mahatma Gandhi understood that 'there are people in the world so hungry that God cannot appear to them except in the form of bread'.

Food is central to our physical and mental health. A sensible diet strongly supports good health, and good health is generally associated with happiness. Despite these links, many of us frequently ignore the importance of a balanced diet in promoting mental well-being. We readily consume unhealthy foods and are riven with guilt afterwards. We live on a see-saw between unrealistic aspirations about what we think we should eat and the dispiriting reality of routine excess. For many people, this roller-coaster is physically unhealthy, mentally exhausting and curiously addictive. Against this background, is it any wonder that so many people are confused and unhappy about food?

There are many underlying reasons for this state of affairs, but much of the fault lies in the exaggerated meanings that we attach to food. Food is far more than a source of nutrition. We commonly treat food as an emotional crutch, a form of communication, a weapon in relationships and even a signifier of who we are. This over-identification with food leads us to associate particular human characteristics with specific foods. It was always thus. The nineteenth-century English art critic John Ruskin wrote that 'cookery means ... English thoroughness, French art, and Arabian hospitality; it means the knowledge of all fruits and herbs and balms and spices; it means carefulness, inventiveness, and watchfulness'. French president Charles de Gaulle asked: 'How can you govern a country which has 246 varieties of cheese?'

This tendency to link personal or national characteristics with food and to overinvest emotionally in our diets leads to excessively complex, unhelpful attitudes towards eating in many parts of the world. We are flooded with books outlining increasingly ludicrous diets, many of which flatly contradict each other and most of which have no evidence to support them. Magazines burst with advice about cooking and eating, and an infinite number of websites offer recipes and meal plans to achieve everything from rock-hard abs to eternal happiness (which are, I emphasise, two separate things).

The end result of this clamour and confusion is that many societies are experiencing problems with both obesity and eating disorders such as anorexia nervosa *at the same time.* While dietary advice that is intended to help this situation can sometimes prove useful, it can also be misinterpreted and inadvertently worsen many of the problems that it sets out to solve. This ubiquity of conflicting guidance causes virtually everyone to obsess unhelpfully about their diet and greatly diminishes our well-being, which is precisely the opposite of what food is supposed to achieve in our lives.

Cultivating a healthy relationship with food is a key step towards overcoming our problems with eating and attaining greater happiness. Let's start with the first problem that besets an increasing number of societies around the world: obesity.

## The problems with food

It is not surprising that we use food for all kinds of things other than nutrition. Comfort eating is easily the most common misuse of food. Many of us use food to deal with emotional discomfort and feelings of emptiness, as if food was a drug (which, in a way, it can be).[1] For this reason, we are drawn to foods that are high in sugar and fat, commonly known as 'comfort food'. We crave these foods when we are sad, lonely, frustrated or angry. We sometimes crave them when we are happy, too, or experience any intense emotion. We always want more. Too much sugar is never enough.

I see this in my clinical work as people use food to try to manage difficult or unstable emotions. I think most of us are familiar with this impulse, at least to a certain extent. Right now, as I type these words, I am acutely aware that there is a chocolate brownie in the fridge downstairs. It is 3 p.m. on a Friday afternoon and I have absolutely no business eating a chocolate brownie at this time. But I'm feeling worn out after a long week. I'm irritated about any number of little things that happened so far today. Even better: nobody else is home, so no one would know if I popped down to the kitchen and did a little comfort eating. Maybe, just this once?

There are several problems with comfort eating in this way, most notably the fact that comfort eating contributes to people becoming overweight or obese. Of course, many other factors are also relevant to our weight, including genes, upbringing, medical conditions, social issues and various other things. But comfort eating is one of the key factors we can control in our day-to-day lives and so increase our health and happiness.

How can we do this? Please bear in mind that the chocolate brownie in the fridge is literally calling to me as I type these words. I can actually hear it.

Let's clarify precisely what we mean when we say a person is overweight or obese. The World Health Organization (WHO) points out that the terms 'overweight' and 'obesity' refer to abnormal or excessive fat accumulation that presents a risk to a person's health.[2] How is this measured? How would I know if I gained excessive weight?

A crude population measure of obesity is the body mass index (BMI), which is calculated using a person's mass or weight, measured in kilograms (kg), and their height, measured in metres (m). The BMI is the body mass divided by the square of the body height (i.e. kilograms/metres squared, or $kg/m^2$). Therefore, a person who weighs 70 kilograms and is 1.8 metres (6 feet) high has a BMI of 21.6. This is a normal BMI. The WHO regards a BMI between 18.5 and 24.9 as normal. A BMI between 25 and 29.9 indicates that a person is overweight. A BMI of 30 or greater indicates obesity.

While these categories are not precise and should be interpreted with care in the context of each individual, they do provide a reasonable estimate of whether a person needs to address a problem with their weight. This is important because being overweight or obese increases the risk of chronic conditions such as diabetes, cardiovascular diseases and cancer. In addition, while overweight and obesity were once regarded as problems only in high-income countries, they are on the increase in low- and middle-income countries too, especially in urban settings. Ultimately, these issues affect every society in the world in one way or another.

Clearly, our increased waistlines have significant consequences for our physical health, but what about mental health and happiness?

There is now growing evidence that, despite many medical and behavioural strategies, the physical and mental health of people who are overweight and obese remain compromised. One research group performed an online survey of 260 people and confirmed that many who are obese have more depression and lower scores on agency, positive affect and flourishing (i.e. engagement, mood and meaning), compared to those with normal weight and those who are overweight but not obese.[3] Clearly, severe obesity is linked with poorer mental health and increased unhappiness.

Another study of Spanish adults with overweight and obesity showed that participants generally perceived themselves as having minor excessive weight, but usually underestimated their actual weight.[4] While most participants tended to report moderate to high body satisfaction, women reported lower body satisfaction and higher weight-related self-stigma than men. Study participants with obesity reported lower body satisfaction and lower self-rated happiness compared to those who were overweight but not obese.

So what does this study tell us? First, these results confirm that there is a relationship between obesity and lower self-rated happiness. Obesity is associated with an increased risk of unhappiness. While this does not apply in every case, it is a general trend. Second, these findings confirm the relevance of weight-related stigma for well-being. Stigma is a powerful force in all areas of life, including weight. Third, the researchers found that positivity is important for the overall well-being of people with excess weight, as people with overweight or obesity with higher positivity traits and body satisfaction are more likely to be happier: having a positive attitude matters greatly to happiness among people who are overweight or obese.

Problems with weight are complex and varied. While obesity lies at one end of the spectrum, people can also suffer from being underweight in the context of disturbed, conflictual feelings and beliefs about food. In some, these problems present in the form of anorexia nervosa, a mental illness characterised by deliberate weight loss induced and/or sustained by the person themselves. In anorexia nervosa body weight is maintained at least 15 per cent below expected body weight or they have a BMI of 17.5 or less (for people aged 16 years or over).[5]

This troubling condition occurs most commonly in adolescent girls and young women but can occur in anyone. The resultant undernutrition can result in various disturbances of bodily function, often relating to specific hormones. The clinical picture is commonly complicated by extremely restricted dietary choices, excessive exercise and further changes in body composition owing to self-induced vomiting and purging, among other behaviours.

Clearly, obesity and anorexia nervosa represent quite different problems with food and weight, but, paradoxically, both conditions also have certain things in common. For example, both obesity and anorexia nervosa are associated with attachment issues, meal-skipping and consumption of fast food.[6] Both are also associated with considerable psychological suffering for those affected by them and for their families and friends.

Obesity and anorexia nervosa are treatable conditions. Obesity is usually managed through lifestyle changes such as adopting a better diet and increasing physical activity. Treatment of eating disorders also involves managing the person's physical health, but with the addition of specialist psychological therapies such as cognitive behaviour therapy, cognitive analytic therapy, interpersonal psychotherapy, focal psychodynamic therapy and family interventions explicitly designed for eating disorders.[7] Medication is sometimes used. A multidisciplinary approach is useful, often involving judicious advice from a dietitian who is familiar with the field. Overall, between 40 and 50 per cent of people with anorexia nervosa recover; up to a further 35 per cent show significant improvement; and approximately 20 per cent develop a chronic disorder, with variable course. In the long term, some people with anorexia nervosa develop osteoporosis or thinning of the bones, predisposing them to fractures. Five per cent die of complications of their disorder.

Obesity and anorexia nervosa are just two of the many problems linked with food and they can result in considerable suffering and unhappiness. There are also other conditions that are, perhaps, not as severe, but still cause widespread unhappiness owing to a poor diet or an unhealthy relationship with food and weight. Many people suffer from a combination of being slightly overweight, feeling chronically guilty about food and experiencing multiple disappointments as diets don't work out or they fail to visit the gym.

For these people, the solution to their problem starts with small, simple changes to their diet along the lines recommended by reliable nutritionists and reputable health services. Most of us love to eat, but if we are to optimise the contribution that eating makes to our happiness, we need to do our best to follow the official guidelines. Let's consider these next.

## A good diet

St Benedict was a Christian saint who lived from around 480 CE until 543 CE and founded 12 communities for monks in Lazio, Italy. Most famously, St Benedict wrote what is now known as the 'Rule of Saint Benedict' in which he presented a set of regulations for his monks to follow. St Benedict's book addresses many areas of monastic life and focuses on values such as humility and trust, simplicity and silence, and listening and living with awe. Judith Valente, in *How to Live: What the Rule of St. Benedict Teaches Us about Happiness, Meaning and Community*, notes that, while the rule covers multiple areas of human activity, St Benedict expresses uneasiness about determining how much a person should eat or drink, regarding this as a personal matter.[8] Nonetheless, he urges moderation in the consumption of wine, which, it seems, he could not persuade his monastic communities to renounce completely.

Nowadays, few people share St Benedict's hesitation about providing dietary advice. In one sense, this is a pity. Much of today's dietary guidance is either dubious or self-serving or both, but happily there is much reliable information to be found – so let's start with Ireland's Health Service Executive (HSE), which is the government agency that provides our public health services in hospitals and communities across the country.

The HSE advises us to eat a wide variety of nourishing foods that supply the energy and nutrients we need to stay healthy.[9] It recommends that we plan our meals in advance so that we introduce variety to our diet and eat more nutritious foods. It suggests using mostly fresh ingredients and snacking on fruit and vegetables. Grilling and steaming are healthier than frying or roasting with oil or fat. Meals should be eaten sitting at a table rather than in front of a television or computer screen.

The HSE suggests that we base our meals on plenty of vegetables, salads and fruits, so that these foods comprise up to half of our plate or bowl at every meal. We should choose wholemeal and wholegrain breads, cereals, pasta and rice, and be aware of the calorie differences between various foods. We should opt for low-fat milk, yoghurt or cheese, prioritising milk and yoghurt over cheese. Our diets can include a small amount of poultry, fish, eggs, nuts, beans or meat at two meals. We should eat fish up to twice a week and limit takeaway and processed food as much as possible.

We are advised to avoid daily consumption of sugary drinks, biscuits, cakes, desserts, chocolate, sweets, processed salty meats (such as sausages, bacon and ham) and salty snacks (such as crisps). Many of these are the comfort foods that we routinely use in our futile efforts to manage our moods with food. But while the short-term sugar rush from the chocolate brownie in the fridge might seem tempting, the official advice is that a sugar habit is unhealthy, hard to kick and not a good way to manage difficult emotions. I'd better leave that brownie untouched for now, I guess. Maybe later?

In the United Kingdom, the NHS notes that most people still do not eat enough fruit and vegetables, which should ideally make up more than a third of our food each day.[10] They suggest that we eat at least five portions of a variety of fruit and vegetables daily, choosing from fresh, frozen, tinned, dried or juiced products. Fruit juice and smoothies should be limited to no more than a combined total of

150 millilitres per day. I routinely break this rule, as I overindulge in orange juice virtually every morning.

Starchy food should make up just over a third of our food intake and we should choose higher-fibre wholegrain varieties, such as wholewheat pasta and brown rice, or just leave the skins on potatoes. Starchy foods are a good source of energy and are the main source of a number of nutrients in our diet. They should not be demonised or ignored.

Milk, cheese, yoghurt and fromage frais are good sources of protein and certain vitamins, and are also an important source of calcium, which is vital for bone health. We should opt for lower-fat and lower-sugar products where possible, such as 1% fat milk, plain low-fat yoghurt or reduced-fat cheese. We should also eat some beans, pulses, fish, eggs, meat and other protein. Pulses such as beans, peas and lentils are good alternatives to meat because they are lower in fat and higher in fibre and protein.

Finally, the NHS advises us to choose lean cuts of meat and mince, and eat less red and processed meat (such as bacon, ham and sausages). Ideally, we should aim for at least two portions of fish every week, one of which should be oily (such as mackerel or salmon). Foods such as chocolate, cakes, biscuits, sugary soft drinks, butter, ghee and ice cream should be eaten less often and in small amounts. Again, I definitely fail here: it is a rare day that biscuits do not make an appearance in my life.

As is apparent, official dietary guidance is consistent across trusted sources such as the HSE and the NHS as well as other national authorities, such as the Centers for Disease Control and Prevention in the United States.[11] Clearly, we should follow this guidance as best we can in our day-to-day lives and encourage other people to do likewise. That way, everyone benefits.

That being said, however, are there specific foods that are particularly good at making us happy, in addition to a generally healthy diet? Is there a magic bullet for happiness to be found on the supermarket shelf? A dietary shortcut to bliss? Maybe.

In 2011 Tyler Graham and Drew Ramsey (a medical doctor) published a book titled *The Happiness Diet: A Nutritional Prescription for a Sharp Brain, Balanced Mood, and Lean, Energized Body.*[12] The title is certainly intriguing: might this book be the pathway to nutritional nirvana?

Graham and Ramsey present much sensible dietary advice placed within the framework of how certain foods can affect our moods. Like much of what we have already discussed, these authors highlight the negative roles of excessive sugar, refined carbohydrates, industrial fats and certain meat and vegetables, depending on how they are produced. Graham and Ramsey also list certain nutrients that have been relatively neglected in our diets in recent times and that can, they contend, increase our happiness. Finally, here they are: the nutritional keys to joy!

First up is vitamin B12, a lack of which increases the risk of irritability, depression and cognitive decline. There is some evidence to support this, so B12, therefore, appears to be particularly important for happiness and can be found in fish, shellfish, liver, beef and eggs. Interestingly, some of these are not foods that most dietary guides suggest we prioritise, but they are important sources of B12, which is essential for the production of brain cells and therefore crucial for our mental well-being. Many of us need more B12.

Other neglected nutrients that can help boost mental well-being include iodine, magnesium and – of all things – cholesterol. Graham and Ramsey point out that dietary cholesterol has a limited effect on our cholesterol levels and is necessary for brain health. Eggs, salmon and meat can all help supply a reasonable and helpful amount of cholesterol in our diets. All told, Graham and Ramsey identify some 12 essential elements for happiness, which are generally neglected in our daily intake.

There is also evidence that omega-3 oils are associated with psychological well-being, along with a range of vitamins and minerals including zinc, magnesium, iron and folate.[13] Omega-3 oils are found in fish, seeds, nuts, olive oil, sunflower oil, avocado and eggs. They may lower the risk of depression and improve mood and might even help protect against dementia. While the effects of these foods are by no means miraculous, it is worth ensuring that they form a part of our diets.

The identification of 'happy foods' does not mean that we should eat large amounts of these foods or exclude other things from our diets, but rather that we should be aware that certain nutrients are essential for brain health, and that brain health is essential for happiness. Even though we cannot literally eat our way to happiness, we can eat our way to unhappiness. And, by correcting our diets, we can significantly increase our physical and mental well-being, and boost our day-to-day happiness.

But how do we actually do this? Eating, like most human behaviours, is overwhelmingly shaped by habit, and habits are difficult to break. Are there any tips or techniques we can use to improve our dietary choices, consume more foods that are conducive to brain health and make ourselves happier in the process?

## Dietary advice

Almost every morning I wake at 6 a.m. and rush downstairs for breakfast. The time at which I wake rarely varies, regardless of whether it is a workday, the weekend or I am on holiday. It is almost always around 6 a.m. and I invariably wake up hungry. Once I have breakfast I am ready to start the day. It baffles me that some people skip breakfast: I have no idea how they function and yet they do. Eating is a habit, so people like me, who are prone to habit-formation, find it difficult to change our dietary practices, especially first thing in the morning.

This tendency to live through habits can make change quite challenging, but it also presents an opportunity; because if I manage to change a habit even slightly, that change is likely to be permanent: if I create a new habit, it will probably be as strong as the old habit it replaces and is therefore likely to persist.

But how do I change my dietary habits to begin with?

The first message is that most diets do not work. They are too extreme. Evolution did not foresee diets at all, because evolution programmes us to consume sugar for an instant energy kick.[14] This served many functions in the past: escaping danger and giving us immediate competitive advantage over other people. As sugary food became more ubiquitous, this evolutionary impulse led us to consume too much of it. Therefore, the extreme dieter has an immediate dilemma as they seek to produce change by suddenly shifting to a different diet, such as completely excluding sugar: evolution is a powerful force.

To compound matters, we are attracted to radical change during moments of epiphany, but dramatic change is rarely sustainable. Rather than revolutionising all aspects of our diets in one fell swoop, it is much better if we make small, incremental changes that we can sustain over time. If I was to abruptly switch from my regular diet to, for example, fruitarianism (a diet made up solely of fruit), it is highly likely that I would fail miserably. Change needs to be planned, gradual and undertaken with the knowledge and understanding of other people in our households. There would be little enthusiasm for fruitarianism in my family circle or wider community.

A great many faddish diets are also highly artificial and do not fit with the rest of our lives. It is difficult to eat like a caveman if you do not live like a caveman. I do not live like a caveman.

This artificiality dooms most extreme diets to failure from the get-go. Mark Twain took this idea to its logical (if absurd) conclusion and remarked that 'part of the secret of success in life is to eat what you like and let the food fight it out inside'. If I were to follow this advice, most of my diet would comprise ice cream, which seems all the more delicious because it is not especially healthy. Voltaire agreed: 'Ice cream is exquisite. What a pity it isn't illegal.' Thank goodness it isn't illegal! Every diet should accommodate ice cream with family or friends on a sunny day. We eat to live.

But Mark Twain had a point. Radical departures from what we are accustomed to eating are unlikely to be successful. 'Intuitive eating' is an alternative approach based on the idea that obsessing over what we eat is bad for our health.[15] Christy Harrison argues that intuitive eating is our body's default mode and that 'diet culture' had led us away from this, to our detriment. Harrison recommends that we rediscover the symptoms of hunger, stop labelling foods as 'good' or 'bad' (most are a mix), become more weight-inclusive in our approach to food and genuinely recognise body diversity. Every person is different – and so are our bodies and dietary needs.

All of this makes sense. So, then, here are the first steps towards improving our eating habits. We should start to recognise when we are hungry. We might be irritable, anxious, unfocused or restless, but these feelings might be indicators of hunger rather than psychological malaise. If so, we should eat something proper. Then, we should figure out when we have eaten enough. We should stop eating at that point. There will be more food later, when we need it again. We should eat what we need when we need it. Then we should stop. This is intuitive eating.

There are, of course, other factors to be considered and these will differ from person to person, depending on what we want from our diets, our bodies and our lives. This is another problem with excessively detailed dietary advice: everyone's needs and priorities are different and likely to change over time. While official dietary advice tends to focus on general health, some people choose to consider other factors. Bodybuilders, for example, commonly adopt incredibly unhealthy diets in order to build muscle mass. This is a conscious choice to prioritise one aspect of diet over another, for a defined reason, even if the resulting diet is highly unbalanced in terms of general health. Biceps like tree trunks are not necessarily healthy.

Sometimes, the best way to change our diet is to focus not on the diet itself but on what we do with our lives and let our diets follow that. Exercise is a good example. If we increase our exercise, we will quickly see that sugary foods do not provide the sustained energy that we need: only a balanced diet can do this. Exercise and dietary change are now routinely recommended as part of treatment programmes for many mental health problems. Even in eating disorders, which are sometimes associated with excessive exercise to lose weight, graded physical activity can prove helpful in combination with other treatments.[16] Often, changes in diet follow changes in activity, rather than the other way around, once we eat only what we truly need.

In addition to linking our diet with exercise, is also useful to repurpose our tendency to attach meaning to food in order to make this habit work for the benefit of our health and happiness. The best example of this is the movement towards growing our own food so as to reconnect with nature and rediscover the sources of what we eat. Too often, food that arrives on our tables is so processed and packaged that the distance from farm to fork feels so vast as to seem infinite. And, yet, each egg was laid by a chicken, each piece of meat was once part of a living animal and every vegetable grew in the soil somewhere. Ultimately, all our food comes from nature.

Reconnecting with the natural origins of our food helps to shape our eating habits in a positive, meaningful direction. Gardening is one of the best ways to make this happen. In *The Well Gardened Mind: Rediscovering Nature in the Modern World*, psychiatriast and psychotherapist Sue Stuart-Smith writes that growing vegetables and fruit using sustainable methods, rather than a highly industrialised food system, allows gardening to provide a bigger story within which people can locate themselves and their lives.[17] Cultivating the earth is empowering and fulfilling. Growing food that you and your family can eat is enormously satisfying and adds a new layer of meaning to our diets, connecting us more firmly with the earth around us.

English physician and writer Thomas Fuller, who lived in the seventeenth and eighteenth centuries, said that 'many things grow in the garden that were never sown there'. This is especially true in relation to the psychological benefits of gardening, which provides not only food and exercise, but also a deep sense of well-being that is often difficult to achieve in other areas of our lives. There is also philosophical joy to be had, as gardening alerts us to the powerful rhythms of the natural world as the seasons pass around us. Stuart-Smith quotes the English poet William Cowper: 'Gardening imparts an organic perspective on the passage of time.'

Growing some of our own food, then, not only supports better eating habits, but also brings us back in touch with the rhythms of nature, the seasonality of foods and the ripeness of what we grow. This matters deeply to our diets, our sense of connectedness with the earth and the meanings that we inevitably attach to food. Ralph Waldo Emerson said that 'there are only ten minutes in the life of a pear when it is perfect to eat'. If you have grown that pear yourself, you will watch for that moment with pride, pleasure and connection. And, when the time is right, you will taste the sweetest pear that has ever graced the earth.

While I lack the space to grow my own food, I have developed a new interest in plants, which is a good start. In fact, the potted plants in my office have now grown so large that it is difficult to travel safely from one side of the room to the other. All the plants are flourishing and spreading beyond my wildest expectations. I wonder do I have the courage to take this to the next level and start to cultivate fruit or vegetables in my office? The science of happiness suggests that I should. Perhaps my long-suffering colleagues might appreciate fresh, office-grown fruit and vegetables for lunch?

## Mindful eating

Mindfulness is everywhere, yet its popularity is easy to understand. Many people feel distracted much of the time: sending text messages, receiving emails, doing video calls and trying to attend to myriad tasks all at once. The practice of mindfulness is the opposite of these things. Mindfulness offers an escape from the hamster wheel of our lives. It appeals to our better selves – and there is no one more seductive than our better selves.

At its simplest, mindfulness means paying attention to the present moment. It involves maintaining a careful awareness of your thoughts, actions and emotions, but not judging them. It means staying focused on the 'now' as much as possible and, when your mind wanders, gently redirecting it back to the present moment: the feeling of the chair you're sitting on, the fragrance of your tea, the person sitting opposite you.

Mindfulness means being present. That simple, direct, unmediated presence is enough. Nothing more is needed. Action is unnecessary. Focus on *now*. For once, don't just do something; sit there.

The idea of mindfulness is rooted in Buddhist and early Hindu psychology and is a key element of meditative practice in both traditions. Over the past few decades, the concept of mindfulness has penetrated deeply into the public mind in many countries. As a result, mindfulness practices now help millions of people in their day-to-day lives and psychological therapies based on mindfulness offer specific benefits for certain people with depression and other mental illnesses.

There is an infinite variety of ways to cultivate mindfulness. One technique is to simply label specific actions or things throughout the day in order to become more aware of them. For example, you could mentally label the position of your body each time it changes: 'walking', 'sitting', 'standing' or 'lying'. This promotes simple bodily awareness, rather than analysis or criticism. The same technique can be applied to thoughts or emotions, labelling each one as it occurs, but not interpreting, changing or lingering on them. See them, accept them, let them go.

The benefits of mindfulness can be subtle but profound. You might still feel irritation at the usual day-to-day problems, but you will be less likely to respond impulsively if you practise mindfulness. You might still feel disappointed at minor upsets, but your disappointment is likely to be proportionate, rather than catastrophic. Best of all, you might discover a new kind of inner stillness, a deepened awareness of the present moment and a reduction in the dark cloud of foreboding that many people report feeling virtually all the time. This can only be good.

How does all of this relate to our eating habits and diets? Can we look to mindfulness to improve our diets and our attitudes towards food, eating and weight? Yes, we can.

A mindful approach to eating means that when we eat, we just eat. At mealtimes, we focus on the process of eating. We avoid distractions and we turn our minds to the act of eating, the taste of food and our body's responses to what we eat. This is both simpler and more challenging than it sounds. There are two main reasons why mindful eating requires some effort before it becomes a habit.

First, as already discussed, we attach a great deal of meaning to food – far more meaning than the food itself merits. Food is not an indicator of who we are as people. It is not a tool to gain advantage in a domestic row or a way to diffuse difficult emotions. Food is just food. Mindful eating means focusing on this thought as we eat, rather than reflecting on the emotions that we wrongly attach to food or the various psychological issues that trouble us in our broader lives, but have nothing to do with food. An apple is just an apple. A steak is just a steak. A biscuit is just a biscuit. They are all innocent, simple, edible. They are not profound or laden with meaning. They are all just *food.*

Second, we tend to set up distractions while we eat, as if we are frightened to focus on the act of eating. I am terrible for this. If I am alone, I find it exceptionally difficult to sit down to a meal without something to read: a newspaper, a magazine or a hardback book that will remain open on its own. Paperbacks are a disaster because they rarely remain open of their own free will and therefore interfere with the act of eating as I need to hold the pages open with one hand. To avoid this, I have balanced bananas, plates, other books and various unlikely household objects on the open pages of paperbacks to keep them open as I eat a sandwich and read at the same time. This is utterly ridiculous. My time is not that valuable.

I know that other people watch television or use laptops while they eat. These are all unhelpful habits because they ensure that we are distracted from both the food we are eating and the book we are reading or the television we are watching. We are concentrating on neither thing properly. This is an error from the dietary point of view because distraction leads to overeating. The solution? When you eat, just eat. When you read, just read. When you watch television, just watch television. This is the essence of mindfulness.

One way to cultivate mindfulness while eating is to focus consciously on the taste of food in your mouth. Henry David Thoreau said that 'he who distinguishes the true savor of his food can never be a glutton; he who does not cannot be otherwise'. Linger over tastes. Notice how flavours emerge and dissipate on your tongue. Reflect on how the tastes feel to you, right here, right now. Take your time with this, focusing on the moment and letting other distractions simply fade away. This takes practice, but it can be done.[18]

Many meditation classes involve mindful eating exercises, such as mindfully eating a raisin.[19] This practice can be difficult to focus on in a group setting, but it is an interesting experience, as you anticipate the raisin, observe it, smell it, touch it, taste it, linger on its taste, reflect on the taste, look back on the experience, and so forth. We are not always in a position to engage in this degree of mindfulness for every bite we eat, but the principle is a good one. We need to be more aware as we eat.

It is, of course, important that we start any mindfulness practice gently and build it up over time. It is good to have a regular pattern and to set aside dedicated time for our efforts. For mindful eating, this might involve eating just one meal per day mindfully to begin with and eating your other meals in the normal fashion. Select the least pressured meal first, so that you have the greatest chance of success. If this does not work out, no problem. Just do your best today and try again tomorrow. You are seeking to build a new habit, which is never easy. Take your time. The speed of your progress is irrelevant. Just be sure to keep going in the right direction, steadily building mindful eating into all your meals. It will become more natural with time.

And if 'mindful eating' sounds too elaborate or unlikely in your life, there are other ways to increase awareness of how our food affects us. One way is to consciously identify and list types of food that elevate our minds and types of food that drag us down.[20] If we undertake this exercise with care, it can make us more aware of the effects of various foods on our mind and increase our appreciation of the role of food in our mental well-being. As far as our minds and bodies are concerned, we really are what we eat.

So if I am what I eat, what exactly am I? Well, most recently, after completing this section of the book, I rewarded myself with that chocolate brownie that I mentioned earlier. I consumed it with no hesitation, great enthusiasm and just enough mindfulness to make sure I appreciated the pure, unadulterated joy of such a wonderful creation. I didn't think about it too much. It just seemed like the right thing to do.

Hippocrates recognised the healing power of good food: 'Let food be thy medicine and medicine be thy food.' This is good advice, but it is not always easy to do the right thing. French author François de la Rochefoucauld wrote that 'to eat is a necessity, but to eat intelligently is an art'. How can we eat intelligently, so as to improve our diets, enhance our nutrition, boost our health and increase our happiness?

The first step is to remember the principles of balance (in our diet), love (for our bodies), acceptance (of ourselves), gratitude (for food and the ability to eat it), avoiding comparisons (with other people's weight or diets) and belief (in our ability to change). This is the foundation for positive progress with our eating habits.

Dietary advice is always a bit vague because each of us is different, with our own needs, goals and issues. St Benedict was wise to refrain from being too specific about how much a monk should eat or drink, regarding this as a personal matter. Psychotherapist Susie Orbach is also correct, in her book *On Eating*, when she advises us to eat what is right for *us*.[21] We need to do this simply, mindfully and with commitment, if we are to harness the power of food to deliver greater health and happiness in our lives.

## Top tips for happiness: diet and eating

- We place too much symbolic importance on food. De-link food from meanings, emotions and self-esteem. See food for precisely what it is: just food.
- Follow official dietary guidance from trusted sources, such as Ireland's HSE, the United Kingdom's NHS and the United States' Centers for Disease Control and Prevention.
- Certain nutrients, such as vitamin B12, can help with brain health. Give them a try, but do not obsess over them.
- A balanced diet will usually deliver all that our bodies need for well-being and happiness.
- Harness the power of habit to improve your eating patterns and rationalise your attitudes towards food and weight.
- Radical diets don't work. They are unrealistic, unsustainable and frequently ridiculous.
- Moderate exercise goes hand in hand with sustainable dietary change as our bodies simply demand the nutrients that we need to keep going.
- Reconnect with the sources of your food, try to grow some of your own supplies and note the seasonality of what you eat.
- Mindfulness can be enormously helpful but does not have all the answers. Mindfulness means being aware of the present moment, simply and directly. Avoid distractions as you prepare and consume your food. When you eat, just eat.
- When you have eaten enough, stop eating.
- Follow your intuition, listen to your body and give thanks for your food and the ability to eat it. It is all treasure.
- Have the occasional chocolate brownie. They are treasure too.

SEVEN

# Move/Stop Moving

Humans are built to move. Movement is essential to health and happiness. And, yet, many of us struggle to exercise as much as we should and a growing minority exercise too much. Some people swing between the two extremes, doing no exercise at all for many months and then bingeing in the gym. We all know that steady habits are best, but we often struggle to attain a balanced exercise pattern in our lives.

I am a perfect example of the problem. On many occasions over the years, I have gone to swimming lessons. I would love to be able to swim. At the pool, I am a diligent student. I listen carefully. I do the exercises. I try my best. Despite this, I repeatedly fail to swim. I have lost count of how many times I have gone through the cycle of resolving to learn to swim, signing up, attending lessons and failing to learn. I do not fear water. I am keen to do better. I see no specific impediment to me swimming like a fish. But, still, I fail. I don't exactly sink, but I certainly do not swim. This is immensely discouraging.

Swimming is by no means my only failure in this field. I have many. Once I joined a gym. I went along on the first day and had an utterly impenetrable 'fitness consultation' with a member of the gym staff. I did not understand a single word he said. He refused to believe that I had never been inside a gym before, saying I must be joking. I wasn't. After around 30 minutes, I thanked him for his time, told him he had been helpful, left the gym and never returned. To this day, I haven't the faintest idea what he was talking about.

The only exercises that worked for me over the years have been gentle ones: hiking and cycling. I do neither activity to a high standard (there is no Lycra involved), but I can hike or cycle at a moderate pace for considerable periods of time. Is this enough? Should I do more? Many people associate exercise with mental well-being. Are they right? Do I need to do more exercise or different activities in order to build my strength, boost my well-being and so increase my happiness?

This chapter focuses on physical activity, on the basis that mental health and physical health are intimately related to each other. But it is also good to 'just sit'; i.e. simply sit in nature. We need to both move and stop moving, consistent with Taoist ideas about achieving greater harmony and balance in our lives. Let's start with some exercise.

## Exercise and happiness

The links between exercise and health have been apparent since earliest history. Hippocrates said that 'if we could give every individual the right amount of nourishment and exercise, not too little and not too much, we would have found the safest way to health'. Socrates agreed: 'No man has the right to be an amateur in the matter of physical training. It is a shame for a man to grow old without seeing the beauty and strength of which his body is capable.'

But what about mental health? Does exercise increase mental well-being and happiness, as well as physical health? President Thomas Jefferson certainly thought so: 'Exercise and application produce order in our affairs, health of body, cheerfulness of mind, and these make us precious to our friends.' One of his successors, John F. Kennedy, agreed: 'Physical fitness is not only one of the most important keys to a healthy body, it is the basis of dynamic and creative intellectual activity.'

Were the two presidents correct? Does the science support a link between exercise and mental well-being and happiness? If so, how strong is the link? And what can we do to optimise our happiness by increasing our physical activity?

There is now extensive research evidence that exercise has significant benefits in depression.[1] One especially compelling study showed that four months of aerobic exercise produces a significant improvement in depressive symptoms and increases happiness.[2] This research did not involve extreme exercise: three 45-minute sessions of walking or jogging at moderate to high intensity each week. The participants in this study were men and women over 50 and the benefits of exercise were undeniable. Exercise alleviates mild to moderate depressive symptoms, improves mental health and boosts overall well-being.

More recent work from virtually every corner of the world confirms this finding. One research group in Iran randomly allocated

120 people, with an average age of 71 years, to either participate in an eight-week, outdoor physical exercise programme three mornings a week or not participate in the programme.[3] Before the intervention, there was no difference in happiness between the two groups, but after the intervention, happiness had improved significantly in the group that took part in the physical exercise programme but not in the other group.

The happiness benefits of exercise are not limited to older adults. One research group in China analysed data pertaining to some 27,706 people over 16 and found that subjective well-being is significantly associated with younger and older age (the U-shape we discussed in Chapter 1), female gender, years of education, better self-evaluated income, self-rated health, social trust, social relationships and physical exercise, among other factors.[4] In terms of physical activity, 78 per cent of people who exercise often report high subjective well-being, compared to 67 per cent of those who never exercise. Interestingly, 77 per cent of those who exercise just sometimes, as opposed to often, report high subjective well-being, indicating that there is little difference between moderate and high levels of exercise in terms of well-being. We do not need to be extreme athletes in order to be happy; we just need to be active.

Another research group, in the Netherlands and Germany, studied 98 people with an average age of 43 to determine the effects of 'Mindful2Work', a six-week programme combining physical activity, yoga and mindfulness meditations targeted at work-related stress complaints from a body–mind perspective.[5] They found that this training produced medium-to-large improvements in key measures such as stress, risk for dropout from work and personal goals. There were also medium-sized improvements in well-being and functioning at work directly after training, six weeks later and six months later, and large improvements after a year. Overall, Mindful2Work showed remarkable, long-lasting improvements in key domains, owing to its combination of physical exercise, yoga and mindfulness. All three of these practices increase physical and mental well-being in different but related ways.

This makes perfect sense to anyone who has experienced the psychological boosts that regular exercise can bring. While it is helpful to see scientific research provide systematic evidence of this effect, the results of these studies should not come as a surprise. Exercise enhances our moods in a significant and lasting way. It is good for body health and brain health and therefore good for happiness.[6]

Does exercise exert its benefits solely by reducing weight? Psychologist Martin Seligman points to links between obesity and diabetes,[7] but also highlights links between fitness and reduced mortality even among people who are overweight.[8] Seligman points out that while most diets are scams, exercise is essentially the opposite of a scam. Between 80 and 95 per cent of people who lose weight on a diet will regain it over the following three to five years, but a much higher percentage of people who take up exercise will stick with it. Put simply, exercise works.

All of these facts point to the wisdom of Plato's view that 'lack of activity destroys the good condition of every human being, while movement and methodical physical exercise save it and preserve it'. But two key questions remain: how much physical exercise do we need and which forms of exercise are best?

Psychiatrist Dr Conor Farren points out that a tiny amount of exercise – gentle walking once or twice a week – is not sufficient to substantially improve our well-being.[9] Research suggests that three sessions of approximately 45 minutes of cardiovascular exercise each week are needed in order to reap substantial mood benefits. Farren describes an interesting case study in which a combination of rugby and relaxation classes helped one man to overcome his recurrent anger problem and regain balance in his life. Similar commitment to physical exercise is needed if we are to feel its rewards.

A similar argument applies to schools. It is insufficient to simply give students an hour in the gym each week, while concentrating sports resources on a small number of high-achieving students.[10] Young people need to be exposed to a variety of sports and activities in order to generate habits of physical exercise that will last a lifetime. Schools provide a unique opportunity to achieve this, yet more could be done to involve every student in sports and physical activities.

Given that most of us already know that exercise is good for us, why do we not do more of it? Why do we remain on the couch, watching television and ordering take-out food? Why are so many of us failing to run, swim or do our yoga regularly?

There are many reasons for this state of affairs, including general inertia, but one of the most recent and paradoxical challenges to sustaining a good exercise pattern is the ubiquity of conflicting advice . Fitness magazines dominate newsstands. An infinite number of online videos and websites offer 'exercise programmes' that are sometimes bizarre or even dangerous. And everyone has their own fitness experience to relate, generally with the punchline that you should do whatever they do simply because they claim it works for them. So what do we need to do in order to be fitter, healthier and happier? How should we exercise?

## How to exercise

The first point is that most of us need to exercise more than we currently do. Humans evolved over many millennia to be active and our bodies crave more exercise in our current sedentary era.[11] We need to stand up, put on our trainers, and head out the front door. But first, some reliable advice about what we need to do as we take our first steps, leaps or hops.

The HSE, the government agency that provides Ireland's public health services, provides detailed guidelines to help us get started and guide us along the way.[12] The key message is that physical activity is for everyone and that any level of activity is better than none. Children and young people (aged 2–18) should be active, at a moderate to vigorous level, for at least 60 minutes each day. This activity should include muscle strengthening, flexibility and bone-strengthening exercises three times a week. Many children are engaged in physical education at school and this is helpful but it is not enough. Some form of training or exercise outside school hours is also needed.

Adults (aged 18–64) should have at least 30 minutes of moderate intensity activity per day, five days a week (or 150 minutes a week). Older adults, aged 65 years or over, should also do at least 30 minutes of moderate intensity activity per day, five days a week (or 150 minutes a week), focusing on aerobic activity, muscle strengthening and balance. A combination of moderate physical activity and muscle-strengthening activity is needed if we are to derive maximum benefits from the time we devote to physical exercise.

In the United Kingdom, the NHS provides advice that is quite similar to the Irish advice.[13] Adults should try to be physically active every day. They recommend strengthening activities that work all the major muscles (legs, hips, back, abdomen, chest, arms and shoulders) on at least two days each week. We should do at least 150 minutes of moderate-intensity activity per week or 75 minutes of vigorous activity.

What is moderate aerobic activity? Moderate physical activity raises your heart rate, makes you breathe faster and makes you feel warmer. If you are doing moderate aerobic activity you can talk but you cannot sing. Examples of moderate activity include riding a bicycle, hiking, brisk walking, dancing, pushing a lawnmower, rollerblading, water aerobics and doubles tennis.

What is vigorous-intensity activity? Vigorous activity means that you are breathing hard and fast. At this activity level, you are not able to say more than a few words without pausing for a breath. Most moderate activities become vigorous when you increase your effort. Examples of vigorous activities include walking up the stairs, swimming fast, running or jogging, riding a bike fast or on hills, martial arts, gymnastics, aerobics, skipping rope and sports such as rugby, hockey and football.

What are very vigorous activities? These are exercises that are performed in short bursts of maximum effort separated by periods of rest. This is also known as 'high-intensity interval training' (HIIT). Examples of very vigorous activities include spinning classes, circuit training, lifting heavy weights, interval running, running upstairs and sprinting up hills. My personal experience with HIIT has been hit-and-miss at best, usually more miss than hit, although I often run up the stairs at some speed for a variety of reasons.

Finally, it is also important to engage in activities to strengthen muscles, which should be done to the point where you need a short rest before repeating the activity. Examples of muscle-strengthening activities include yoga, t'ai chi, Pilates, carrying heavy shopping bags, lifting weights, doing exercises that use your own body weight (e.g. sit-ups, push-ups), working with resistance bands, heavy gardening (e.g. digging, shovelling), lifting and carrying children or wheeling a wheelchair. In my experience, yoga is an especially useful way to strengthen muscles, as it combines physical movement with mindfulness in a pleasing combination that is as enjoyable as it is healthy.

In addition to these activities, it is helpful to reduce the time we spend sitting or lying down and to break up long periods of not moving with some activity. The weekly activity target outlined by the NHS can be met with several short sessions of very vigorous intensity activity or a combination of moderate, vigorous and very vigorous activity. There is also more specific exercise guidance for children, young people and older adults, all focused on the benefits of regular physical exercise.[14] In the United States, the Centers for Disease Control and Prevention offer similar advice and excellent day-by-day guides to assist with exercise.[15]

So if we are not in the habit of exercising regularly, where do we begin? How do we embark on a sustainable programme of physical activity in order to make ourselves healthier and happier?

Dr Mark Rowe advises that we should have a medical check-up before commencing a moderately vigorous exercise programme, warm up, start gently, go slowly and always stay hydrated.[16] In terms of specific activities, we tend to engage in team sports when we are younger, but move towards individual sports as we grow older (walking, swimming, jogging).[17] The key is to remain engaged in some form of physical activity, no matter what our age. Signing up for a gym or pool that is several miles away is unlikely to produce a positive outcome. We tend to find ingenious reasons to avoid exercise, especially as we start out, so it is important that physical activity fits neatly into our daily routine. Exercise is supposed to reduce stress, not increase it. Overall, we were born to be physically active, moving makes us happy and increases our resilience, and physical activity helps us manage stress.[18] It can help if exercise has a social dimension.

Many people find that technology helps them set targets and monitor progress, although others find that technology gets in the way of zoning out of the world when they are walking or cycling. I find that technology gets in the way of mindful exercise, which involves focusing on the present moment while we run, swim, cycle or do whatever exercise we choose. It is important that we take mindful care of our whole bodies through regular exercise, especially if we have back problems.[19]

A reflective exercise such as a body scan can help you get in touch with your body before you exercise and increase awareness as you move. As we discussed earlier, a body scan involves focusing your attention at the top of your head and then moving down your body, noting each part as you progress and consciously relaxing as you go along. It is important to be systematic, mindful and slow. A body scan is an excellent way to 'think your way into your body' before you warm up or exercise.

We need to exercise most days each week, with a combination of moderate to vigorous physical activity and strength exercises. The weekly goal of 150 minutes is achievable for most adults once we design our exercise plan, commit to doing our best, tolerate our failures and always try again tomorrow, no matter how unsuccessful we are today. Here it is useful to apply the six principles of a happy life discussed in Chapter 3: balance (between rest and activity), love (especially self-love), acceptance (of certain limitations), gratitude (for what we can do), avoiding comparisons (gyms are full of improbably fit people) and belief (in our ability to improve).

In recent years, many people have chosen running as their exercise of choice, with the result that running has now achieved near-mythic status in the pantheon of physical activities available to us. So let's consider running in more detail. Just how miraculous is it?

## Is running special?

I repeatedly try to get into the habit of running and repeatedly I fail. It does not help that I intensely dislike running. I can think of few things I enjoy less, to be honest, so presumably my efforts to run are doomed unless I have a radical change of heart. This is a pity, because running is commonly hailed as a magical cure for every psychological problem imaginable. Newspapers gush about the spiritual uplift of a good jog. Websites offer running diaries and inspirational quotes about the mysticism of jogging. Apps invite us record our progress and respond to automated messages of encouragement telling us that we are doing great (even if we are not). It is all a bit wearisome, to be honest, and it certainly does not encourage me to run.

All told, running is now held up as the apogee of human existence, a sort of physical elixir for all ills. But is it? What is the evidence for this? And are there any alternatives to running for those of us who fail to see even the faintest attraction in going for a jog? Are we doomed?

Isabel Hardman, in her book *The Natural Health Service: What the Great Outdoors Can Do for Your Mind*, provides an excellent account of both the enthusiasm of runners for their sport and their predilection for expounding the apparent psychological benefits of running to others, often at great length.[20] Hardman notes the weary nods of non-runners who try to tolerate gushing descriptions of the joys of running, when they have always found the activity perfectly hideous whenever they tried it. Even so, Hardman outlines clearly the physical and mental health benefits that can be associated with running. But while she makes a compelling case in favour of the sport, there is still an unanswered question about whether or not running is *extra* special, over and above other forms of exercise. Is the *particular* enthusiasm of runners for their chosen activity entirely justified?

In fairness, there is extensive research to show that physical movement improves our mood and, for some people with mild depression, running can be just as effective as psychotherapy.[21] Enthusiasts also describe 'runner's high' which occurs when endorphins and enkephalins are released just as the runner is about to be overcome by exhaustion. This helps the runner to run beyond the pain in a state of elation, it seems. While I have never experienced this, I have heard it described many times by breathless runners keen to convert others to their cause. Of course, having a passing sense of euphoria, although pleasant, does not necessarily improve mood in the longer term. Only an exercise *habit* can do that. Still, transient euphoria is probably a good start.

Hardman, in her book, discusses 'parkruns' at some length and these are an excellent example of how a sport such as running can become a healthy habit. Parkruns are free, weekly, timed events that take place across the world, organised by local volunteers.[22] The concept is simple: you turn up at the appointed place and time on a Saturday. You then walk, jog or run five kilometres or, if you are a junior, two kilometres on Sunday. It does not matter what you wear or how fast or slow you go. Taking part is what matters. Everyone is welcome. You can also volunteer in various roles at parkruns. The focus is on inclusiveness, participation and well-being. Founded by Paul Sinton-Hewitt CBE in London in 2004, parkrun's mission is to create a healthier, happier planet. As far as I'm concerned, parkrun now makes a significant contribution to this rather ambitious task.

Parkruns are incredibly popular. They are free, outdoors and healthy, and the distance of five kilometres is an attainable goal for a majority of people. Parkruns are also social events at which runners of every speed are made welcome. So just how helpful are parkruns? They sound great, but is there systematic evidence that they improve our physical and mental health?

As the parkrun phenomenon spread around the world, it inevitably caught the interest of researchers. One research group asked new parkrun registrants to complete self-reported measures of physical activity, weight, happiness and stress, at time of first registration with parkrun, 6 months and 12 months later.[23] They obtained additional information for their analysis from the parkrun database. In the end, 354 participants completed the study.

The results of this analysis showed that levels of physical activity increased significantly over the first 6 months following registration with parkrun and, by 12 months, weekly physical activity was 39 minutes higher than at the start. There were also significant reductions in BMI over 12 months and weight loss of 1.1 per cent in the whole sample, rising to 2.4 per cent among overweight participants. Even more interestingly, there were modest increases in happiness and decreases in perceived stress, along with an overall 12 per cent improvement in fitness. All told, the physical and mental health of people who participated in parkruns improved significantly, even one year later.

Making exercise into a habit is the key to success. Parkrun is a great way of doing this with running. Not only do we tend to stick with our habits, but there is evidence that the benefits of exercise are greater if exercise is habit rather than something we do at random intervals. One research group performed brain scans on 40 people and found that the reduction in anxiety produced by exercise was linked with the person's level of habitual physical activity: participants with a high level of habitual physical activity showed significant reductions in anxiety after exercise.[24] Clearly, we should try to make exercise into a habit if we are to enjoy its benefits to the fullest extent possible.

So, armed with this knowledge, and with several parkrun locations available near my home, why do I still fail to run? I think the problem lies in part with my personal history (which does not involve running), the hours I have spent listening to runners extolling the virtues of running (which cause me to tune out) and the simple difficulty I experience with changing my deeply ingrained habits (which do not include sufficient physical activity).

Perhaps I need to reconsider my position on running, now that I am faced with growing evidence that going for a regular run brings so much benefit to so many people. There is certainly no doubt that jogging, running and being involved in running clubs *can* help with physical and mental health.[25] Therefore, if something along these lines works for you, it might well bring all kinds of personal and social benefits in the long term. For other people, however, including me, it might be that running simply does not suit our needs, preferences, temperaments or lifestyles. If this is the case, there are many other exercises that can readily deliver similar benefits.

Cycling is an excellent alternative to running. Cycling need not involve expensive racing bikes, embarrassing Lycra outfits or high-speed sprints on urban streets. Cycling has the merits of being relatively easy on our joints, building muscle and providing an aerobic workout that benefits our hearts, blood vessels and brains.[26] Cycling also helps to build bone, which is especially important as we age, and has spill-over benefits for walking, standing, balance, endurance and climbing stairs. As with all sport, it is important to cycle safely, seeking medical advice beforehand if needed. It is also advisable to wear a helmet, ride with someone else, use bike paths rather than streets, stay hydrated and use sunscreen and sunglasses, as indicated.

If activities like running, cycling or swimming do not meet your needs, regular brisk walking is another a good option, once it is sufficiently vigorous to induce perspiration. Henry David Thoreau said that 'an early-morning walk is a blessing for the whole day'. He was absolutely right. Thomas Jefferson agreed: 'Walking is the best possible exercise. Habituate yourself to walk very far.' Ralph Waldo Emerson points that 'few people know how to take a walk. The qualifications are endurance, plain clothes, old shoes, an eye for nature, good humour, vast curiosity, good speech, good silence and nothing too much.'

Regardless of which sport or activity we choose, the key is that we engage in moderately vigorous physical exercise, three times a week, for around 50 minutes each time (or equivalent activities differently distributed). We should do our chosen activities regularly, combine them with strength exercises, and try to make the experience as enjoyable as possible. For me, that mostly means cycling, although it is likely that I will resume my doomed efforts to swim from time to time over the coming years. Who knows? I might even try running again. Anything is possible.

## Sitting still

The world of exercise is beset with image problems for the uninitiated. Gyms, fitness studios and similar venues commonly advertise with pictures of impossibly fit-looking, smiling people performing unlikely tasks: lifting enormous weights, contorting their bodies in gruesome gym machines or running long distances in improbably beautiful locations. Most potential customers know that no matter how much they train, they will never look like the people in these images. Some might find such pictures inspiring, but I venture that a majority (including me) find them off-putting and emblematic of a deeper problem in the world of modern fitness: smugness, exclusivity and a self-righteous attitude that pretends to invite all comers, but visibly exults in its own perceived superiority.

It was always thus. In the first century BCE, the Roman statesman Marcus Tullius Cicero advised that 'it is exercise alone that supports the spirits, and keeps the mind in vigour'. I agree with Cicero that exercise supports good mental health and happiness, but I do not agree that 'exercise alone' does so. As we have seen, many other things matter too: age, genetic inheritance, childhood experiences, family relationships, earning money, being employed, physical health, religion, political beliefs, where we live, how we sleep, how much we dream and what we eat. Exercise is important, but it is certainly not the only thing we need.

Despite these facts, Cicero was far from alone in deifying the place of physical activity in our lives. We see this constantly in newspaper features, blogs and websites which intimate that running or some other form of exercise holds the solution to all of life's ills. But while exercise is both necessary for health and conducive to happiness, physical activity is certainly not the only requirement for well-being and it does not merit its recent promotion to fill the gap left by religion in many people's lives.

As ever, the Buddha offers a balanced perspective, stating that 'to keep the body in good health is a duty ... otherwise we shall not be able to keep our mind strong and clear'. Exercise matters, but other things are needed too, including a clear mind, which the Buddha suggests is the purpose of good health in the first place. Hippocrates highlights the need to couple exercise with a good diet: 'Even when all is known, the care of a man is not yet complete, because eating alone will not keep a man well; he must also take exercise. For food and exercise, while possessing opposite qualities, yet work together to produce health.' In other words, many things are needed for health and happiness, so exercise is just one part of a much bigger jigsaw puzzle. In a sense, that is what this book is all about: a wide-angle picture of the path to happiness.

One of the activities that has been most neglected in relation to happiness, possibly as a result of the exuberance surrounding exercise, is not an activity at all: sitting still. I am exceptionally enthusiastic about sitting still as a vital component of any programme aimed at increasing our mental and physical well-being. I am not alone in this view.

The nineteenth-century physician William Osler said that 'patients should have rest, food, fresh air, and exercise – the quadrangle of health'. It is interesting that Osler put 'rest' in first place on his list. Osler was one of the brightest medical minds of his generation and possibly one of the finest ever. The emphasis he placed on rest was undoubtedly deliberate and, in my view, entirely appropriate. It is especially important to articulate the importance of rest today, when sitting still is an increasing problem for many people who struggle with distraction, excessive activity, obsessional exercise or simple difficulty tuning out of work and social media for any appreciable period of time.

This is a pity. Sitting still is just as important as exercising if we are to attain the Taoist sense of balance as one of the six principles of a happy life (the others being love, acceptance, gratitude, avoiding comparisons and believing). If we are serious about achieving balance in our lives, keeping our bodies and minds still is just as important as moving them. Humans might have been built to move, but movement only has meaning when it is interspersed with periods of stillness, calm and simple being rather than doing. The next chapter of this book is explicitly devoted to doing and not doing, and will explore the value of meditation, a particular form of sitting still, as an antidote to stress and excess. What I am talking about here is not so much meditation as rest: simply doing nothing.

It is tempting to stay on the move all the time; some people even develop exercise addiction. To combat these tendencies, we need to stop, calm down and rest. When people think of resting, they often suggest an activity to help with it: gentle walking, watching television or even going for a massage. These activities can certainly help to a point, but they are still *activities* that can become stressful in their own ways. We can be irritated if interrupted while watching television and some people simply cannot tolerate massages: I myself am repelled by the very idea and the one time I agreed to try a massage, I fled the building before the poor bewildered masseuse came within a metre of me.

No, I am talking about rest, pure and simple.[27] Stepping back and doing nothing. This is a skill that is different from laziness, different from 'active relaxation' and different from what most of us are accustomed to. But periods of rest are highly restorative for our bodies and minds, beaten and bruised as they are by the punishing lifestyles we lead. I frequently prescribe rest to my patients, especially those who are worn out by the infinite demands of life. Rest is magic.

To return to Cicero: 'Exercise and temperance can preserve something of our early strength even in old age.' This balance between activity and temperance is one of the forgotten keys to well-being. While we increasingly (and correctly) sing about the joys of physical exercise, we fail to appreciate the quiet satisfactions of temperance, the replenishment of rest, and the simple, thrilling magic of doing nothing at all.

This chapter was all about moving and stopping moving, exercising and sitting still, doing and being. We need a healthy balance of all these activities if we are to increase the likelihood of happiness in our lives. In a world of distractions, it is tempting to stay on the move all the time: doing rather than being. This can apply to excessive exercise, as well as compulsive work. As with all activities, injudicious or obsessional exercise can become an addiction and lead to health problems. While there is little risk of that with me, it is a real problem for many people who simply cannot stop.

So just as we need to learn to exercise, we need to stop exercising. And as much as we need to learn to move, we also need to sit still. Just that: just sit.

## Top tips for happiness: exercise, moving and stopping moving

- Regular exercise enhances physical health, increases mental well-being and boosts our happiness.
- Humans were designed to move, so we function best with regular exercise.
- Most of us need to do more exercise.
- Adults need to be physically active every day: do at least 150 minutes of moderate intensity activity or 75 minutes of vigorous activity each week.
- Fit in some strengthening activities that work all your major muscles (legs, hips, back, abdomen, chest, arms and shoulders) two or three days each week.
- For many people, running is the ideal activity for physical and mental health.
- For other people, different activities better meet their needs: cycling, swimming, brisk walking, other sports, any vigorous activities or even energetic forms of dance (although the less energetically I dance, the better).
- The key to establishing a regular exercise habit is that our chosen activities should be sustainable, convenient, enjoyable and (ideally) sociable.
- Parkrun fulfils many of these requirements, is widely available, and free.
- Finally, it is important that physical exercise is coupled with an ability to sit still, a vital but neglected part of the exercise cycle.

EIGHT

# Do/Stop Doing

We are busy people. Our heads are full of thoughts, plans, ideas, worries and a million forms of self-doubt. We are coming, going, staying, doing and wondering what other people think about us. We struggle to sit still. We are filled with inner restlessness. We try 'active resting' in order to fulfil our conflicting needs for rest and activity at the same time. It does not work. We respond to nervous energy by creating more of it. This does not work either.

And so our lives remain busy, our minds remain cluttered and we feel that we have not achieved enough. A lot done, a lot more to do. There is no finish line. The wheel keeps turning for ever.

This is an unsustainable situation that makes us all unhappy. How can we remedy this problem in order to achieve a healthy balance between activity and rest, movement and stillness, doing and not doing? Is a solution even possible?

Perhaps the first step is to acknowledge the problem. We need to own up to the perils of constant busyness. It leads nowhere. Socrates advised us to 'beware the barrenness of a busy life'. He was right. American inventor Thomas Edison agreed: 'Being busy does not always mean real work. The object of all work is production or accomplishment and to either of these ends there must be forethought, system, planning, intelligence, and honest purpose, as well as perspiration. Seeming to do is not doing.'

We often hear that being busy is not the same as being productive. This is true, but the tyranny of productivity is little better than the tyranny of busyness. Edison's emphasis on 'accomplishment' as well as 'production' is important. We need to feel *personal* accomplishment and reward for what we do, not just the excitement of *doing*. Leisure time matters too, as does opportunity for rest and reflection. If we do not have enough of this, our busyness lacks purpose or reward. We lose sight of what matters in a whirl of futile, misdirected activity. Aristotle believed that leisure was the purpose of busyness in the first place: 'Happiness is thought to depend on leisure; for we are busy that we may have leisure, and make war that we may live in peace.'

Increasingly, we neglect the leisure side of this equation and spend our lives at war with an unspoken fear of quietude, silence and stillness. We find it easier to keep going than to stop. Why?

It is often said that if a shark stops swimming, it will die. This is a myth. Even if it was true, humans are not sharks. If we stop, we do not die. We live. The ability to sit still is a skill that we have systematically devalued, subsequently lost and need to regain if we are to achieve greater happiness.

## Clutter everywhere

Like many people, I get a buzz from being busy. I fill my diary to bursting point each day. I fly from task to task. I am happy when everything is done.

To keep track of my busyness, my diary functions as a to-do list. I keep my old-fashioned paper diary with me at all times. When I complete each task or finish each appointment, I draw a line through the relevant diary entry. Once I have put lines through everything in my diary for the day, I go home. While some days are busier than others, each day always has a list of tasks that I check off as I do them. Often the satisfaction I get from ticking tasks off the list greatly exceeds the satisfaction I glean from the tasks themselves. My life is ruled by lists.

This is not necessarily a good thing. While the daily to-do list gives me both control and the feeling of control, it also implies that my life is simply a succession of tasks that grows longer as I grow busier. The system accords little value to the process of doing things, only to the fact of having them done. The system also fuels busyness, as the to-do list gets ever longer. Am I, like many people, not only busy, but *pointlessly* busy and possibly inefficient as a result? Or, worse again, pointlessly productive? Does this constant stream of activity generate greater achievement or does it, as Socrates suggested, lead to a barren life, as I rush from task to task, rarely stopping for reflection? And where does happiness fit in to this picture?

There is a real issue here. Many people struggle to accept life as it is and this can result in extraordinary levels of activity.[1] Scientist Francis Crick agrees with the value of avoiding unnecessary busyness: 'A busy life is a wasted life.' For many, happiness has taken second place to busyness. This is a pity, especially when our busyness is not essential. We attend meetings that are utterly pointless or efficient only for the person who called the meeting. For everyone else, the meeting is likely to be a poor use of time. We perform many tasks for the sole reason that we have always done them or we imagine someone else wants us to do them, when, in fact, they might prefer if we left matters alone. In short, we waste time frantically, at great length and in ingenious ways.

In a struggle to cope, many people develop unhelpful habits or even addictions.[2] One of the key ways to remedy this problem is to identify what, precisely, we do with our time. It is surprising how little attention we devote to this question. Henry David Thoreau said that 'it is not enough to be busy. So are the ants. The question is: what are we busy about?' This question is important for many reasons, not least because we often struggle to identify which of our activities are worthwhile and which are not. Ralph Waldo Emerson said that 'we do not know today whether we are busy or idle. In times when we thought ourselves indolent, we have discovered afterward that much was accomplished and much was begun in us.'

In a nutshell, we are far too busy and, for much of the time, we do not have the faintest idea of the value of what we are doing. One of the psychological reasons for this state of affairs is that we have let busyness become a symbol of virtue and accomplishment. We regard the busy person as admirable and have less time for the person who says that they often leave work early because all their work is done. And yet, the second person might well be the more organised, efficient, productive and happier of the two.

There are many reasons why we value busyness in ourselves: a desire to feel essential, a need to contribute to society, avoidance of (imagined) criticism and low self-confidence.[3] Busyness is also rewarded and reinforced from earliest childhood and this value system carries over into adulthood as we implicitly link activity with self-worth. We end up as 'human doings' rather than 'human beings'. We dash between tasks, scarcely pausing to breathe, let alone evaluate where we are going or what we are doing. We simply keep moving, no matter what.

Often, it is not the quantity of work that creates this sense of busy panic, but the sheer complexity of our lives and the infinite number of small actions that steal our time and energy. It is the clutter of tasks that is the problem, rather than the amount of work that needs to be done or that we (mysteriously) feel we need to do. So how can we fix this?

Once we have diagnosed the problem, the next step is to find a way to declutter our lives and minds, which is just as important as decluttering our homes (something that has become an obsession in recent years). To return to Thoreau: 'Our life is frittered away by detail. Simplify, simplify.' We can achieve this goal by performing just one task at a time, setting aside past and future concerns as we do so, and focusing on the present moment. This is easier said than done, but it is essential that we try. The benefits are enormous.

Let's start with decluttering.

## Decluttering

Simplicity lies at the heart of decluttering. It links with our need to reconnect with life's simpler pleasures and become less reactionary, more forgiving, kinder and more willing to consider other points of view.[4] It all sounds wonderful. How do we begin?

In recent times, decluttering our homes is usually the first step recommended to start this kind of process of renewal. The annual clear-out and clean-up that was previously known as 'spring cleaning' has been re-branded as 'decluttering' and marketed as a way to transform our lives as well as our homes. Notwithstanding the breathless hyperbole surrounding decluttering, it remains the case that decluttering has much to recommend it, just like spring cleaning did (with considerably less fanfare).

Marie Kondo is the undisputed queen of decluttering. Kondo is a Japanese organising consultant, author and television show host.[5] The fundamentals of Kondo's approach are familiar to the many millions of people who have read her books, watched her on television or come across any of the countless articles and discussions about her work over the past decade. Kondo's website provides a pithy summary of her Six Rules of Tidying, which involve committing yourself to tidying up, imagining the lifestyle you desire, discarding things first, tidying by category rather than location, following the correct order and asking yourself if each item sparks joy.[6] According to Kondo, her KonMari Method can make your home permanently tidy and clutter-free, with the result that your whole life will change: you will be more confident, more successful and motivated to build the life that you truly want. Her advice is surprisingly practical and well worth a look.

But what if decluttering our homes does not transform our lives? Kondo's approach might help tidy the kitchen, but does it help us to create the life we would like to lead? Can we apply Kondo's principles more broadly and tidy our way to happiness?

This approach can help in the world of work as we apply the principles of tidying to our workspaces, digital work, time, decisions, networks, meetings and teams.[7] The two recurring themes are the value of decluttering in all these domains and the need to focus on activities that spark joy. This is based on the idea that joy at work sparks joy in life, which is largely true.

Focusing on activities that spark joy does not, however, mean that we abandon everything we do because none of it makes us immediately happy. Decluttering our diaries does not mean that we simply avoid all activity. Exercise is linked with both physical health and happiness, and there is also evidence that non-exercise activity makes us happy too.

One research group used a smartphone application to collect self-reports of happiness and physical activity from 10,889 people and also gathered information about physical activity from accelerometers on participants' phones.[8] The findings were clear: people who are more physically active are happier, and they are happier while they are more physically active. As a result, the researchers conclude that not only is exercise associated with happiness, but other physical activity is too. On this basis, we should not seek to simply remove activity, especially physical activity, from our lives in a blunt attempt to declutter, but we should be more selective in the activities that we choose to pursue, ensuring that they make us feel joyful or contribute meaningfully to goals that matter to us.

Sarah Reynolds, a professional organiser, provides useful guidance in her excellent, no-nonsense book, *Organised: Simple Ways to Declutter Your House, Your Schedule and Your Mind*.[9] Reynolds suggests that we create a list of short- and medium-term things to do, divided into home tasks and work tasks. She recommends managing paperwork in such a way as to facilitate retrieval, ensure we know what paperwork we have and why, allow access to important paperwork as needed and permit the development of our interests and skills. Paper itself is not necessarily a problem, but how we manage it can be. It should not hold us back in our efforts to create joy in our lives and achieve our goals.

Decluttering our home and work environments will usually help us to clarify our objectives, streamline our activities and hopefully eliminate much of our pointless busyness. The key to decluttering our activity schedule lies in performing just one task at a time and having a clear understanding of why we are doing each activity, what the consequences will be and what would happen if we did not do that particular task. Just how necessary is it? If we decide to go ahead, we need to set aside past and future concerns, focus on the present moment, commit to the task at hand and find joy in the activity as best we can.

Selecting tasks can be a challenge. If individual tasks do not spark immediate joy, they might nonetheless spark joy later as part of a broader effort aligned with a medium- or long-term goal. Curating our activities therefore depends not only on the immediate joyfulness of each task, but also on the extent to which we are truly motivated to achieve our goals: reaching the correct balance between doing and not doing requires an awareness of what we are trying to do in the first place and what motivates us. Let's consider this next.

## Motivation

I am standing on one foot in a public park in Dublin. My left foot is on the grass. My right knee is bent. My right foot is raised and pressed against the inner thigh of my left leg. My hands are in front of my chest, pressed together as if in prayer or as if I am about to say 'Namaste' – a Sanskrit phrase that means 'I bow to you.' In fact, I am not about to say anything. I am at an outdoor yoga class and

I am about to fall, having wobbled dangerously for the past few moments. It is not easy to balance on one foot in tree pose. I feel too tall for this. I routinely fall over. Today is no different. My stance is precarious at best and possibly a risk to public safety. Should I really be doing this in public?

But then my yoga instructor – an impossibly kind, bendy person – reminds us: 'Find your *drishti*.' *Drishti* is a Sanskrit term that means focused gaze or way of developing concentrated intention. In this yoga class, 'finding your *drishti*' means finding a spot straight ahead of you and staring at it. Focusing the gaze in this way helps with balance. I duly select a spot on a tree trunk at the edge of the park and stare at it fixedly. A few seconds later, I stop wobbling. I will not fall. I have found my *drishti*.

Finding our focus is central to committing to any project that we choose to undertake.[10] Focusing on motivation steadies us up and keeps us on track. Unfortunately, it is easy to become distracted from our goals and, as a result, we often fail to appreciate our own motivations, let alone those of other people. This is a pity. Understanding motivation is useful for deciding which tasks we should pursue and which we should skip, and it is a good way to declutter our lives in a mindful, positive fashion.

Many people believe, not without reason, that humans are fundamentally motivated by the desire for money and greater wealth. This is true to a significant extent: a certain amount of income assists with happiness, although the benefits of additional income diminish significantly beyond a certain point. So if the happiness generated by income is limited, what motivates us to work beyond that point? Why do we continue? Is it a desire for power? The admiration of others? Habit?

In *Drive: The Surprising Truth About What Motivates Us*, Daniel H. Pink points to the importance of intrinsic motivation, which is the inherent satisfaction that we sometimes find in certain activities we choose to do.[11] This kind of motivation is self-directed and depends on three key elements: autonomy, mastery and purpose. Pink argues that we need to shift from extrinsic to intrinsic motivation if we wish to strengthen our organisations and address the lingering feeling that something has gone wrong with our lives and in our worlds.

Pink makes a good point that is strongly supported by research in this field. We need to focus more on satisfaction in our work (not just earnings), cohesion in our communities (not just wealth) and well-being in our societies (not just economic output). In 2008 French president Nicolas Sarkozy indicated that he was dissatisfied with the state of statistical information about the economy and society, and he established a Commission on the Measurement of Economic Performance and Social Progress in order to refocus public policy on happiness and well-being.[12] The commission made a range of recommendations for an improved approach to evaluating economic performance and social progress, emphasising both objective and subjective dimensions of well-being, social inequality and environmental sustainability. In essence, the commission recommended broadening the canvas when we think about well-being in our societies.

The importance of the new approach was duly reflected in leading media outlets such as the *Financial Times, Economist* and *Irish Times*, and similar initiatives were commenced elsewhere.[13] In 2010 the United Kingdom launched the comprehensive Measuring National Well-being programme (MNW). In 2019 New Zealand announced that it would introduce a national well-being budget.[14]

All of these initiatives reflect the facts that human motivation is more complex than previously imagined and factors such as purpose and service help to sustain us in our work and build better, stronger communities.

How can we use this information to select which tasks to prioritise in our day-to-day lives and which to omit? Pink, at the end of his book, suggests 20 conversation starters that can help trigger a dialogue about motivation with our friends and colleagues. He draws particular attention to the idea of purpose, asking how much of our activity has true purpose, be it in our work life, family life or other areas of activity. Purpose is vital. Refocusing on purpose might mean changing the balance of our activities at work, shifting emphasis within our family life or seeking out voluntary activities that add to the sense of meaning in our lives.

All of these steps help, but many people struggle with them. In my work as a psychiatrist, I commonly see people whose day-to-day problems diminish their ability to see beyond the immediate tasks that confront them. This is entirely understandable. In these situations, it is useful to refocus on the six principles of a happy life: balance (between daily necessities and medium-term goals), love (especially maintaining self-love), acceptance (of both the requirements of the day and the need for a greater vision), gratitude (for the ability to both attend to today and imagine tomorrow), avoiding comparisons (other people's successes are rarely what they seem) and belief (in goals that seem distant but can still inspire us). Helping others and connecting with friends can remind us of the broader context in which we live and rebalance many of our excessive worries.

In addition, we saw that religion and spirituality contribute to individual happiness and good mental health, despite the associations between certain religions and concepts such as guilt and intolerance. Politics is also relevant to happiness. Taking all of this together, it is clear that having set of beliefs about how the world should be, and our place within it, can help to clarify our purpose in life and chart a path forward. Once we focus on a goal that motivates us, we can achieve great things. We can separate the essential from the inessential. We can prioritise activities that make us truly happy and eliminate much of the pointless busyness that dominates so much of our lives at present.

To do this, we need to understand our motivation, declutter our lives and learn to focus. As my yoga teacher says, we need to find our *drishti*. But how, in a world of haste and frenzy, can we find the mental space to do this?

## Meditation

In 1660 French scientist Blaise Pascal pointed to the perils of busyness and to a possible solution in his *Pensées*: 'I have often said that the sole cause of man's unhappiness is that he does not know how to stay quietly in his room.' Staying quietly in our rooms is certainly one way to avoid the busyness that obsesses us so much of the time. But what are we to *do* in our rooms? Just sit there? How does that help with purpose, motivation and focus?

Meditation is one of the best ways that we can spend time in our rooms, simply *being* rather than constantly doing. Meditation is especially useful because it compels us to notice just how busy our minds and bodies usually are. It requires us to calm our frenetic thoughts so that we live more fully in the present moment, rather than rushing heedlessly into the next one. Meditation is not about disconnecting from the world, but connecting more deeply with the world *right now* and disconnecting from the imaginary worlds that occupy us for so much of the time. How do we begin?

The best way to start a meditation practice is to commit to meditating today *and only today.*[15] Longer-term commitments tend to stop us before we start, so just decide that you will meditate today. That is plenty for now. Do not worry about tomorrow.

Next, find a relatively quiet spot in which to sit. You will never find complete silence and you will never be entirely certain that nobody will disturb you. There are noises and people everywhere. Just find the best place that you can and, if you are disturbed, notice the disturbance as it occurs and then watch it dissipate of its own accord. You do not need to do anything. Once the distraction has faded, guide yourself gently back into your meditation.

It is not necessary to sit cross-legged. Take up whatever position is most comfortable for you, ideally either sitting or lying on the floor. If you tend to fall asleep when you lie down, then sit up. As you sit, make sure your spine is relatively straight. Place your hands in front of you, possibly on your knees or the arm-rests of a chair. Your feet should be planted firmly on the ground and your gaze more or less straight ahead or slightly downward.

Try to focus on your breathing. This is difficult: once you direct your thoughts to your breath, a million other thoughts will enter your head: events at work today, things you have to do later, engagements coming up tomorrow, issues from the distant past, random thoughts that you didn't even know were in your head, thoughts about thoughts, thoughts about thoughts about thoughts, and so forth. We carry a multicoloured infinity of ideas in our minds. All of them come rushing in as soon as we try to focus on our breath. So be it!

Do not be dismayed by this. If we were able to focus clearly on our breathing, we would not need to meditate. Our minds would already be disciplined, calm and directable. We would be Zen masters.

Instead, our minds are like troupes of monkeys chattering and jumping all over the place with no apparent pattern or purpose. Noticing this is the first and possibly most important step in meditation. Our monkey minds are immensely complex, powerful and filled with energy, so if we learn the discipline of contemplative practice, our minds can achieve great things.

Rather than focusing simply on the breath, many people find it helpful to structure their focus in a particular way. One common technique is known as the 'mindfulness of breathing', which we mentioned in Chapter 4. To do this, count your in-breaths for ten breaths, so that you have counted from one to ten over the course of ten in-breaths. Then, count ten out-breaths. Finally, count ten turnings of the breath (after the in-breath and before the out-breath). Then, start again. This is a common meditation exercise that helps us to relax at night-time and, during the day, connects us more fully with our breath, our bodies, the present moment and the worlds within and around us.

Meditation sounds like a simple exercise, but staying focused is a challenge. It is important that we persist. Making time is vital, as are self-compassion and self-love. It is inevitable that we will be distracted, but it is not inevitable that we will be disheartened. When distractions occur, notice them, try not to respond and watch as they pass out of awareness. Refocus on your breath. Meditation is a practice, so imperfection is part of the deal. If we do not need to refocus our thoughts, then we are not doing it right. Bringing back our wandering minds is the work of meditation. Without that, we achieve little.

What are the benefits of meditation? At one level, the purpose of meditation is to help us to see reality as it truly is. To begin with, this is the reality of the present moment while we sit and meditate. In time, this mindset becomes a habit and, eventually, it becomes our default position, even when we are not meditating: being rooted in the moment and seeing the world as it really is. This approach helps us to remain calm, connected and aware. It settles our thoughts, boosts our mood and promotes lasting well-being.

These are big claims, so it is easy to be cynical about mindfulness and meditation. In fact, it is vital that we question the benefits of any practice, such as mindfulness or meditation, in order to be certain that we are using our time well and not deluding ourselves with seductive worldviews that deliver little benefit in our lives. Happily, an increasing body of research supports the usefulness of mindfulness.[16] There is now strong evidence from randomised controlled trials to indicate that mindfulness interventions (especially eight-week mindfulness programmes) help with the management of chronic pain, reduce rates of relapse of depression in at-risk individuals and improve outcomes in substance abuse.[17]

Despite these benefits, there is little doubt that mindfulness and the wellness industry in general have been oversold as instant cures for all life's ills in recent years.[18] It is, however, important that excessive claims made about mindfulness and meditation do not draw attention away from the proven benefits of these ancient practices. As is the case with many things, mindfulness and meditation do not have all the answers, but they are valuable tools when they are used with humility, wisdom and care.

We are busy people who need to slow down. This is difficult. Busyness is addictive. It makes us feel valued and important. Even so, it is possible to achieve a better balance between doing and not doing in our lives, and so boost our happiness.

Decluttering our lives will increase our well-being and make us happier. How we go about this varies from person to person, but, in general, we should be aware that we overvalue busyness and action and we undervalue leisure and inaction. Aristotle was right: 'Leisure of itself gives pleasure and happiness and enjoyment of life, which are experienced, not by the busy man, but by those who have leisure.'

To experience this, we need to slow down. And sometimes we need to stop.

## Top tips for happiness: doing and stopping doing

- Too often, we equate work with productivity, and activity with value. Try to dismantle these beliefs.
- Declutter today. Our lives are filled with unnecessary tasks, unfounded beliefs and physical objects that weigh us down rather than move us forward.
- Identify the objects and activities that bring you happiness and focus on these.
- Perform this task with an awareness of your motivations. Do your current activities align with a sense of purpose in your life?
- Self-direct your activities when possible and prioritise activities that deliver a sense of inherent satisfaction.
- Focus on creating mental space to maintain a calm focus on what matters.

- Meditation helps. Learning to sit still in our own presence is an important life skill. Try it. Create the time. Practise.
- Specific techniques such as mindfulness do not hold all the answers, but are useful as part of a broader commitment to more contemplative, happier living.
- Sometimes, less is more – and even less is even more.

NINE

# Connect/Disconnect

Several days each week, I go to a bakery to buy bread. I know the staff by name and they know me. They nod when I come in. I like that. The staff predict what I want, even though my order changes slightly from time to time. I like that too. Hector, who is often behind the counter, offers an observation about the state of the world, a thought for the day or – on special occasions – a joke he has saved up. Hector's jokes centre on the absurdity of human nature. They are gentle and generous, delivered with perfect timing. Hector's insights and humour elevate each bakery visit into something that cheers me up for the entire day. 'So much more than bread,' Hector muses, as I leave the bakery, happy.

Connections matter. Everybody counts: family, friends, colleagues, clients, patients, casual acquaintances and – perhaps especially

– Hector in the bakery. Together, all of these people create the complex mosaic of humanity that shapes our lives. We are part of their mosaics, too, and the lives of many others, in continually overlapping circles that stretch to infinity. We are all connected.

These connections to other people serve many purposes. They help us through the bad times. Aristotle said that 'the antidote for fifty enemies is one friend'. President Woodrow Wilson agreed: 'Friendship is the only cement that will ever hold the world together.' An earlier president, Abraham Lincoln, asked: 'Do I not destroy my enemies when I make them my friends?'

Connections with family and friends amplify the good times and challenge us to do better, do more or do less, as circumstances require. We need the different perspectives of those around us if we are to thrive and grow. The virtue of friends is that they are like us, but not us. Plutarch understood this: 'I don't need a friend who changes when I change and who nods when I nod; my shadow does that much better.'

On the flip side, relationships can bring much that is unhelpful, unhealthy and even unbearable into our lives. People upset us, ignore us and walk away. It is hard not to take this personally. It is difficult not to attribute someone else's bad behaviour to our own perceived failings. This is because our family and friends matter so deeply to us. We care about their opinions. As a result, they hold unique powers to disappoint, anger and upset us.

Against this background, it is clear that managing our connections with other people is central to our happiness. Humans are social creatures. We need each other, but we also need to ensure that we do not make each other unhappy. With this in mind, this chapter focuses on connecting with, and disconnecting from, other people. Research shows that connecting with happy people creates happy networks and makes us happier, but disconnecting is equally important. Technology plays both positive and negative roles, so we need to be mindful of how we use smartphones and social media in our relationships. Connections are about people, not phones.

Let's begin by looking at the most fundamental question of all. Do relationships really make us happy? What is the evidence for this?

## Does connecting make us happier?

In 1930 British philosopher Bertrand Russell wrote *The Conquest of Happiness*.[1] Russell starts his reflections by discussing common causes of unhappiness, including many that we have touched on here: competition, boredom, excitement, fatigue, envy, a sense of sin, persecution mania and fear of public opinion. Russell is particularly strident about the evils of envy, which he rates as probably second only to worry as a cause of unhappiness. This is similar to the idea of avoiding comparisons as one of the six principles of a happy life.

Having considered the roots of unhappiness, Russell turns his penetrating gaze to causes of happiness, including zest, affection, family, work, interests, effort and resignation. Resignation is similar to 'acceptance', another one of the principles of a happy life. Russell advises against fretting over every detail and points out that we waste much time and emotion by overinvesting in minor matters. We need to accept life's smaller troubles and keep them in perspective. There are always bigger fish to fry.

Many of Russell's views on happiness and unhappiness are careful, considered and surprisingly contemporary, and much of what he says is supported by the happiness research of recent decades. But it is his views about connecting with other people that are of greatest interest and value to our present discussion.

In the section about the family as a cause of happiness, Russell laments the fact that the potential of the family to increase our well-being and satisfaction is not always realised in practice. This idea appears to be a constant feature of human thought: that the institution of the family is in a state of disarray at best and disintegration at worst. Despite continual predictions of doom, however, the family continues to prove one of our most enduring social arrangements. It is subject to change over time, but it is usually expanded rather than contracted, as the concept of family is redefined and extended to include differently constituted groupings. Despite all the complaining and lamentation, we appear to hold an unshakeable belief that family relationships increase our well-being and meet our emotional needs. This is probably mostly true.

Russell's consideration of affection more generally as a cause of happiness is just as interesting as his discussion of the family. He writes that the best type of affection is reciprocally life-giving, such that each person gives affection without effort and receives it with joy. This is the ideal relationship and it contributes greatly to happiness, but even Russell acknowledges the existence of other kinds of relationships, in which the giving and receiving are severely unbalanced, causing unhappiness for all. Managing these relationships is complex but not impossible.

More recent literature highlights the role of relationships in boosting happiness, maintaining psychological well-being and improving physical health.[2] Good relationships help us to flourish in all respects and are among the greatest gifts that life can bestow.[3] They can, however, also open us up to loss, emotional vulnerability, disappointment, anger and a range of other complex emotions. It's a tricky, rewarding balance.

Perhaps the strongest evidence of the importance of relationships for happiness across communities comes from the Framingham Heart Study, the long-term follow-up study we discussed in Chapter 1.[4] This study found that clusters of happy and unhappy people were clearly visible within the social network examined. People who were surrounded by many happy people were more likely to become happy in the future, as happiness literally spread through the network. Clearly, a deeper understanding of social networks is vital if we are to better navigate the challenges facing our world.[5]

All of this is consistent with most people's gut feeling that good relationships matter greatly and with recent research that reveals the exquisitely social nature of the human mind. Humans were designed to live in groups.[6] The functional biology of the brain, insofar as it is currently known, strongly supports the idea of a social brain that is primed for connection and love: no matter which way you look at it, relationships matter deeply on emotional, psychological and biological levels. We were made to mingle.

Relationships also matter therapeutically, helping us to deal with psychological distress and the crises that life invariably sends our way. This is, perhaps, one of the most valuable features of human relationships: they evolve over time and thus provide a unique bulwark against the random tides of human affairs. Most of us know this from our own lives, as we look to family and friends for companionship, consolation, hope and – in Hector's case – bread.

With all of this in mind, how can we purposely build better relationships, so that we give and receive affection in the way that Russell describes? Are there specific steps we can take to better connect with others?

## How to connect

The first step in connecting better with other people is connecting better with ourselves. Most of us need to improve our relationship with ourselves before we reach out to others. This sounds logical when we think about it, but, increasingly, people do not make the time to think about these issues or to reflect on their lives. Too often, we simply rush from task to task, problem to problem, without deepening our awareness of how we see ourselves and how we refer to ourselves within our own minds. We let our inner critic run free. We fail to reflect on our lives in a calm, realistic and constructive way. We do not accept ourselves as we are.

To remedy this, Dr Harry Barry recommends that we move away from the ideas of self-esteem and self-rating, and replace these with unconditional self-acceptance.[7] This means accepting ourselves as we really are and taking responsibility for our behaviour. This is good advice: the idea of 'self-esteem' is an exceptionally unhelpful one, with its implications of measuring, valuing and judging. We simply cannot 'rate' any human being, including ourselves.

I see this problem often. People with anxiety and depression are commonly plagued by poor self-esteem and, as a result, look for specific reasons to think better about themselves. Yet it should not be necessary to find such reasons. We should think highly of ourselves by virtue of the simple fact of being human. We should not need specific reasons to accept ourselves as we are, to value ourselves and to love ourselves.

Building self-acceptance involves facing reality, accepting it calmly and then deepening self-compassion. Instead of asking if we are good enough, we should ask: What is good for us?[8] Rather than rating ourselves and evaluating our 'self-esteem', we should focus on self-kindness, self-compassion and our common humanity. This does not mean that we never seek to change our lives for the better, but rather that we undertake projects such as dietary improvement and more exercise from a place of self-care rather than self-criticism.

Once we accept ourselves as we are, we have a much stronger basis for relationships with other people. The next step is to review and renew the relationships we already have, rather than rushing out to make new ones. We are all embedded in family and social networks already, even if we have let many of the threads become distant and frayed. Refreshing and renewing existing relationships is an important first step in widening our social circle and reaching out to friends, family, friends of family, friends of friends and the broader world.

When we move beyond these circles, sports, book clubs and social events all provide connections that can begin online but need to develop into in-person friendships if they are to fulfil our emotional needs. Pets help, especially dogs that need walking and cats that need to be rescued from neighbours' gardens. I especially recommend cats for meeting people. Cats are weirdly hypnotic.

While all of the resulting social relationships matter deeply, it is often romantic relationships that occupy much our thinking about connecting with others. For people who are already in a romantic relationship, refreshing this connection can take many different forms. Making time to be together and communicating clearly is vital. Henry David Thoreau said that 'the greatest compliment that was ever paid me was when someone asked me what I thought, and attended to my answer'. Listening is the key to being heard, and listening lies at the heart of any long-term relationship, such as marriage.

Many other things are needed too, to keep a marriage strong. For example, it is helpful to create some unpredictable moments and unexpected pleasures in our relationships in order to revive the element of surprise that characterises the early stages of many romances.[9] Novelty is an underrated virtue, even in a 40-year marriage. Small gestures mean a lot.

For those who are still in search of love, it is useful to recall the advice of philosopher Friedrich Nietzsche that 'it is not a lack of love, but a lack of friendship that makes unhappy marriages'. Friendship is vital if romance is to grow. The world of online dating has commodified the way that many people meet, but once the initial connection is made, the relationship is still an interpersonal one that requires both friendship and romance if it is to thrive. In one sense, dating apps have changed everything, except everything that matters, and the rest is still up to us.

Romantic love has hijacked the broader concept of love, so it is important that our rush to romance does not distract us from other relationships with family, friends, colleagues and acquaintances. It is important that we feel that we are part of some kind bigger organism: a family, club, company, school or similar grouping.[10] In 1999 Irish psychiatrist Anthony Clare described involvement with this kind of group as being like a leaf on a tree.[11] We are all happier when we feel part of something bigger, when we are leaves on trees and when we recognise that the enormous value of community groups and activities lies in deepening relationships and increasing mutual happiness.[12]

Making and keeping friends is a skill that richly rewards our efforts and adds greatly to our well-being. A small number of close friends can be more valuable than a large circle of distant ones. Aristotle again: 'He who hath many friends hath none.' A close friend offers things that family members cannot: new perspectives, objective opinions and someone to whom you can complain about your family. Greek tragedian Euripides expressed this well: 'One loyal friend is worth ten thousand relatives.'

The keys to friendship are similar to the essential elements of any other relationship: communication, affection and loyalty. Socrates recommended being 'slow to fall into friendship; but when thou art in, continue firm and constant'. Most of us can count on one or maybe two hands the family members and friends on whom we can depend for anything – people in whom we confide and whom we could contact at 4 a.m. to help us out of a scrape. These relationships are worth their weight in gold. And, like gold, they tend to make us happy.

## Disconnecting

So far we have focused on the benefits of human connectedness and the joys of relationships of various kinds: romantic ties, family relations, friendships and casual acquaintances. But there can be significant downsides to relationships: violations of trust, jealousy, envy and gradual distancing. Even good relationships can prove intense and emotionally demanding at times. As a result, disconnecting from other people is just as important as connecting with them in the first place. Everyone needs a break and, ideally, we should all spend time alone, away from other people.

The problem is that disconnecting can be just as difficult as connecting, if not more so. Humans are primed to value social information, personal messages and, above all else, gossip. Stepping away from this is a real challenge, especially if we are heavy users of social media as our chief means of communicating with the world.

There have been many studies of the positives and negatives of social media and our reliance on devices to connect with others. Results are mixed, but most good studies demonstrate two key points: spending too much time on screens is not good for us, but technology holds much that is positive as well as negative. Let's look at some of the negatives first.

In 2018 one research group looked in detail at decreases in psychological well-being among American adolescents and possible links to screen time during the rise of smartphone technology.[13] They examined data from 1.1 million adolescents in the United States between 1991 and 2016. They found that adolescents who spend more time on electronic communication and screens (e.g. social media and the Internet) and less time on non-screen activities (e.g. in-person social interaction, sport) have lower psychological well-being. Adolescents who spend a small amount of time on electronic communication are the happiest.

This finding fits with much of the other research on this theme and confirms what many people suspect anyway: too much screen time is a bad thing. But does screen time cause unhappiness, or does unhappiness lead to increased screen time, as people seek to address their psychological problems by spending more time online? Intriguingly, the research group found that, for the most part, changes in levels of online activities *precede* declines in well-being, suggesting that increased screen time is followed by a fall in well-being, rather than the other way around. While this does not definitively prove that high levels of screen time *cause* unhappiness, it provides convincing evidence that increased time spent on electronic rather than in-person communication tends to be followed by a decrease in happiness.

So what are the upsides of using electronic communication to connect with others – or are there only downsides?

It turns out that there are many upsides, starting with the short-term kick of hearing about the emotional lives of minor celebrities or seeing your neighbour's holiday photos. If these delights wear thin, it is worth remembering that smartphones and social media provide extraordinary access to people's lives and therefore a unique opportunity for positive interventions to make people happier. There are many studies of apps and other interventions aimed at boosting psychological well-being. One study looked at a smartphone intervention to cultivate hope as a way of promoting well-being.[14] Initial results are positive, suggesting that further research is justified and that we can use our virtual connections in a positive way.

Despite these benefits, it remains the case that most of us need to work harder on disconnecting from our devices, stepping back from our relationships and taking time away from our day-to-day lives in order to be alone. This is not always easy, especially if you have a busy family life, work in a competitive environment (in which those who are always available do best) or are addicted to social media. But these factors make disconnecting more important, rather than less. We all need to take a step back.

When you decide to disconnect a little, it is important to be realistic. Most of us cannot, and do not want to, walk away from our lives and relationships, in order to live in the desert for a year, with just a camel, the sand and our thoughts for company. We need to retain some links with our worlds, but also find ways to set them aside for periods of time. How we do this varies from person to person, but going for a walk or cycle for an hour without your phone is always simple and cost-effective. Going to the cinema or theatre alone is another way to disconnect and enter a different world for a couple of hours. So too is a night away, a weekend retreat or an evening class, such as yoga, language lessons or learning how to type.

The key is that we should create a time and space just for ourselves, in which we are unconcerned about our relationships with other people and essentially uncontactable. This can be difficult for parents of small children, but it is important that we try as best we can, whenever opportunity arises. (For parents, the message is that whenever anyone offers to assist with child-minding, the answer should always be Yes.)

Of course, most of us will never achieve the supreme degree of disconnectedness that a monk attains in a mountain-top hermitage in the Himalayas. Complete disconnection from the world is simply not possible.[15] As a result, our failure to find perfect solitude should not prevent us from taking advantage of the short periods of relative aloneness that we can achieve in the context of our lives.

We need to be alone. As much as we need to connect with others, we also need to disconnect. We should do this frequently, regularly and without the slightest hint of guilt.

## Connecting in the time of Covid-19

In early 2020 the sense of balance that we seek in our lives and relationships came under unprecedented pressure with the Covid-19 pandemic. As the virus spread from China around the globe, our personal and public worlds changed in profound and unpredictable ways. We were advised to remain physically distant from each other, stay at home as much as possible and wear face coverings in certain settings. Suddenly, our environments were more dangerous and unknowable than before. The future was uncertain.

To compound matters, communicating with other people became a particular challenge. More communication moved online, which brought certain benefits but was also distancing. Face coverings, essential to prevent transmission of the virus, diminished the richness of human contact. Seeing someone's unmasked face became a moment of intimacy.[16] Within a few weeks, we discovered how much we rely on lip-reading and facial expression to understand each other. Not only was there anxiety and increased needs to connect, but there was also confusion and slippage of meaning, despite our desire to reach out.

Coping with these circumstances is difficult.[17] Often, relationships sustain us in troubled times, but with Covid-19 other people became the greatest threat of all, because they might transmit the virus. The relationships that structured our lives up to that point were undermined by distance, uncertainty and fear. Friendships became strained, family members were physically distant and anxiety was everywhere. How can we survive emotionally in circumstances such as these, when our brains are fundamentally wired to connect with other humans?

Ultimately, despite the fact that the virus is transmitted from person to person, it was soon clear that relationships would lie at the heart of our coping strategies. People found ways to meet up online and, when possible, in person, subject to physical distancing and face coverings. Governments around the world published detailed guidelines that permitted certain activities once we took specific precautions. Public health officials put their heads together to figure out which gatherings presented high risks and which were less likely to transmit the virus. Schools were cautiously reopened so that children could attend classes and socialise as much as possible, subject to public health guidance.

This was a time when friendships, connections and relationships became more important than ever, as we adjusted to the strange new world of Covid-19. To return to Euripides: 'Friends show their love in times of trouble, not in happiness.' Suddenly there was a new balance to be struck between connection and disconnection as the rules shifted towards online communication and away from face-to-face contact. Negotiating the new landscape was not easy, but sheer necessity forced everyone to look at new ways of connecting with family, friends, colleagues and even potential partners. We still reached out, just in different ways.

The key to much of this lay in recognising the centrality of relationships in supporting each other, accepting the new rules of engagement and valuing in-person meetings more than ever, once appropriate precautions were observed. Humans are adaptable and while the 'new normal' did not seem 'normal' to any significant degree, it became workable for many people who sought new means to deepen and refresh their relationships. In circumstances such as these, relationships matter more than ever, so we should prioritise connections above all else: more effort is needed, but the rewards are greater too.

Many of the challenges that people reported as the pandemic emerged were related to anxiety, fear and, often, despair. This is a pity. Notwithstanding the pandemic, past decades present myriad reasons to be broadly optimistic about the future. Overall, our lives are becoming longer, healthier, safer and more prosperous.[18] Our understanding of human nature is also evolving, challenging the idea that humans are by nature selfish and governed by self-interest, and focusing on trust and cooperation instead.[19] Our better angels are stronger than we think and might even lead to a solution for the climate crisis in the end.

These ideas were largely upheld by the response to Covid-19. Communities came together. People sang from their windows during quarantine. Communities grew stronger than ever.[20] Despite the pressure, relationships held, connections deepened and we clapped for front-line workers. While many of the people I saw struggled with both anxiety about Covid-19 and the effects of public health restrictions, they also spoke about mutual support, community initiatives and the value of reaching out to others. These things do not happen on their own. They require investment in relationships and the kinds of robust connections that have sustained human beings for centuries – and will sustain us for centuries to come.

Connections matter. Relationships with other people are an integral part of what it means to be human. Our brains are wired to connect with each other. Trust is central to good relationships. In the words of Epicurus: 'It is not so much our friends' help that helps us, as the confidence of their help.'

Research confirms what most of us already know: good relationships assist with physical and mental health and hold unique potential to make us happy. As a result, many of us overinvest in particular relationships. We become obsessed with the idea that the perfect relationship will solve all our problems. It won't.

Happiness depends on many things. What matters is that we belong to a happy *network*, within which happiness spreads from person to person. This reflects many of the principles of a happy life that we discussed in Chapter 3: love (especially self-love), acceptance (of other people), gratitude (for the relationships in our lives), avoiding comparisons (which erode relationships) and belief (in the fundamental goodness of people, including ourselves).

In the end, we make each other happy just as much as we make ourselves happy.

## Top tips for happiness: connecting with and disconnecting from other people

- The first step to connecting better with other people is connecting better with ourselves. Take time to become more aware of your thoughts, moods and emotions. Reflect. Keeping a journal can help.
- The second step is to deepen, enrich and renew the relationships you already have with your partner, family members, friends, colleagues and acquaintances. Reach out.
- Remain open to accidental friendships that might occur at work, socially, in the gym, as you walk the dog or – best of all – while you search for your cat.
- Online dating has changed the way that many romantic relationships begin, but such relationships need to move to the in-person stage if they are to meet our emotional needs.
- Being part of something bigger matters deeply, be it a family, a book club, the company where you work, the school you attend, a political movement or any similar grouping. A sense of belonging nourishes us, even if we never explicitly draw on the support that it offers.
- We sometimes need to disconnect and take time alone. This can be for short periods, such as evening walks without the phone, or longer periods, like weekend retreats. Just go.
- Achieving the optimal balance between connecting and disconnecting can be a challenge, especially in stressful times, but it is worth making an effort to achieve this balance. We will all be happier as a result.

TEN

# Lose Yourself

We have lost the art of getting lost. As adults, we tend to know where we are all the time and can predict with a good degree of certainty where we will be tomorrow, next week and next year. This is regrettable. Getting lost is an important part of living.

The experience of getting lost changes as we move through life, mainly for the worse. Virtually all children get lost from time to time and most know that their parents or guardians will find them soon enough. For those who do not enjoy secure attachment to an adult, the fear of getting lost can be profound. But most children are happy to wander about, secure in the knowledge that, in the end, they will be found.

As we become adults, we lose this sense of becoming 'securely lost'. Most commonly, we slip out of the habit of getting lost altogether and we maintain a continual, detailed awareness of our location at any given time. We live overdetermined lives in a hypercognitive world. We forget how to loosen our moorings, drift away and spend time neither knowing nor caring about our coordinates on a map.

This chapter examines the idea that it is particularly difficult to get lost in the modern world owing to technology, addiction to information and an inability to let go. We will look at various instances of me getting lost in such exotic locations as Tokyo, St Petersburg and Bangalore. These are all places in which it was relatively easy to become disoriented, lose myself in my surroundings and let go of conscious awareness of where I was, what time it was and where I was going next. As it turned out, these were important experiences. We overcontrol our lives. We need to let go.

## Japan: found in translation

My flight from Heathrow to Tokyo in the early 1990s was the first long-distance flight of my life. I travelled with a group of other 'young Europeans' on the first leg of a two-week tour of Japan organised by the Ministry of Foreign Affairs of Japan. I was one of two representatives from Ireland. We met the others in Heathrow. They came from countries all over Europe including Denmark, Portugal, France, Spain, Italy and the United Kingdom. We were giddy with excitement; a bunch of adolescents and young adults setting out on the trip of a lifetime. The Japanese spared no expense: we flew first class on Virgin Atlantic and spent two glorious weeks in Tokyo, Kyoto, Hiroshima and various other places. It was a magical trip.

I was already in love with Japan long before this journey. Some months earlier, in the latter part of my teens, I had written an essay about the films of Akira Kurosawa to earn a place in the group, drawing links between the Japanese director's work and the theme of the Foreign Ministry's travel programme: 'Tasks and Future Perspectives of the Relationship between Japan and Europe.' My essay touched on Japanese and European theatre, politics and literature, as well as the prints of artist Hokusai:

> So too did Vincent van Gogh aspire to the directness of Japanese art. He recognised in these prints the freedom which was denied to him as a result of the inhibited European tradition into which he was born. Indeed, writing to his brother, Theo, in 1889 about his masterly 'Artist's Room in Arles', he commented: 'The broad lines of the furniture, again, must express absolute rest ... I shall work at it again all day, but you see how simple the conception is. The shading and the cast shadows are suppressed, it is painted in free flat washes like the Japanese prints ...' It was in these Japanese prints that van Gogh finally found the liberty, simplicity and directness to which he had so long aspired.

I had a memorable interview at the Japanese embassy in Dublin prior to the journey. My interviewer asked about various aspects of my essay, which I duly discussed. He then said that when he meditates, he thinks of the moment at the start of the universe and the moment at the end of the universe. With that, he fell completely silent and looked at me expectantly. I nodded, smiled and said nothing, wondering if he could read my mind. I was always relaxed with silence, so I just sat there. After a good five minutes, my interviewer smiled, mysteriously satisfied with my wordlessness. I was on my way.

Arriving in Tokyo was one of the most sublime experiences of my life; and when I stepped out of Tokyo's New Otani Inn on my first afternoon there, I felt I had landed on another planet.

Within minutes I was lost in the streets of Tokyo: completely, utterly and deliriously lost. This was the era before the Internet or mobile phones. I was on my own with just a paper map in my pocket, but the map was no use to me: the streets were unidentifiable, the directions unfathomable. There were crowds of people everywhere, countless restaurants, neon lights and strange noises coming from all angles: music from tiny bars, laughter from amusement arcades and a million unidentifiable sounds from the thronging streets. It was intoxicating. I felt a little dizzy, as if I were collapsing into chaos.

But the infinite swirl of people, sights and sounds was anything but chaotic. This was a carefully choreographed scene, immaculately managed and utterly hypnotic. A sea of pedestrians surged across a traffic intersection and suddenly stopped dead at the very instant the traffic lights changed, frozen in obedience. Hostesses outside bars beckoned me in with beautiful smiles, gesturing at menus and bowing sweetly. Schoolchildren glanced at me shyly, their eyes wide as they beheld a ungainly, six-foot foreigner standing awkwardly in the middle of the footpath as masses of people flowed smoothly around him.

I was lost – utterly at sea, without the faintest idea how to find my way back to my hotel. But while the Japanese street scenes I beheld were truly amazing, perhaps the most remarkable aspect of my situation was how relaxed I felt. I was thousands of miles from home, alone in an unfamiliar city, profoundly dislocated and without any way of remedying the situation, yet I felt strangely reassured and not the slightest bit worried. Somehow, I knew that all would be well.

Afternoon became evening, evening became night. Even in the darkness, munching noodles at a street stall, I felt entirely at ease, unworried and strangely at home. Why? This was most unlike me.

I have thought about this over the years and have reached the conclusion that I was what I now call 'securely lost'. This means that while I did not know where I was, I was reassured that nothing bad would happen. I would be found. Or, rather, I would find my way home. Somehow, I knew that everything would be fine, so I could just enjoy my dislocation for what it was: a simple, transient joy.

Even so, my relaxed attitude still makes little sense because the foreignness of Japan was just as profound and overwhelming as it was intoxicating. I had literally no idea what was going on around me. And yet, I felt reassured, possibly by a combination of the orderliness of what I saw (the streetscape ran like clockwork), the fact that Japan is a relatively safe country (with one of the lowest crime rates in the world) and that I was on the adventure of a lifetime.

Japan is also famous for inducing a sense of unreality and dislocation. Ten years later I revisited this feeling when I watched and rewatched the movie *Lost in Translation*, written and directed by Sofia Coppola. This film is set in Tokyo, where actors Bill Murray and Scarlett Johansson convey the feelings of cultural displacement and belonging that Japan can so readily invoke, as Tokyo both alienates and draws us in, soothing us with its hypnotic strangeness and beauty. The film, like Tokyo, is mesmeric.

I eventually arrived back at my hotel and drifted up to my room to sleep for 14 hours. The next day, I wondered if the entire episode had been a dream, but I revisited many of the streets over the following week and recognised them at once. The sights and sounds continued to fascinate me, but now that I was with the group, these scenes did not hold quite the same allure as they previously had. I was no longer lost.

I had a wonderful fortnight in Japan. We visited shrines in Tokyo, temples in Kyoto and the Hiroshima Peace Memorial. We stayed with Japanese families, I went to a baseball game and the tour organisers scheduled activities to suit the interests of each of us. I had written much of my essay about Japanese film, so, at one point the organisers took me aside and made a deep, deep apology (as only the Japanese can), saying that Kurosawa was filming in the north of Japan and would be unable to meet me. Would it be acceptable if they brought me to a film studio instead, to meet other people from the film industry, including a cameraman who worked with Yasujirō Ozu, the other giant of Japanese cinema? Yes, I said, trembling with excitement, that would be fine.[1] The studio visit was a dream come true.

But while the Japanese trip was remarkable in all respects, nothing quite matched the subversive strangeness of being lost on my own in Tokyo on that first evening, the joy of being utterly adrift and the thrill of feeling the world as something new, unknown and without limits. In one sense, much of my life since then has been spent trying to regain that feeling, trying to be securely lost in a world that increasingly lacks the kind of strangeness and dislocation I found so readily in Japan.

The experience of being securely lost requires both a feeling of firm attachment that allows us to loosen our moorings in the first place and the ability to let go of the moorings and drift away a little when the situation is right. For me, Tokyo in the early 1990s unexpectedly met both of these requirements. A similar experience is more difficult today owing to the ubiquity of technology, especially mobile phones, and our addiction to information about where we are in the world. This is a pity. Getting lost and being found make us happy. We need a balance of both in our lives.

## Russia: a night in St Petersburg

In previous chapters we have focused on systematic, scientific research into happiness, centred on things that we measure, like exercise and well-being, or income and happiness. These parameters are all important and – to an extent – quantifiable.

But what about things that cannot be measured and still influence our happiness? The happiness I felt getting lost in Japan could not be predicted, measured or studied in a systematic way. How can random events and circumstances like these be built into our view of happiness so as to inform our paths towards greater well-being?

The idea of attachment is central to the way we incorporate these kinds of incidental happenings into our lives. If a child is securely attached to a parent or guardian, the child is happy to wander off within a certain distance, try things out, make mistakes and still know that all will be well. They will be rescued. The insecure child becomes hysterical as soon as their parent or guardian leaves their line of sight. They hesitate to take risks. Every mishap is a catastrophe. This pattern echoes into adulthood.

Mistakes are vital for well-being. Sometimes, the greatest happiness is when things go wrong and we end up on an unexpected road, when a failure in one respect is an unexpected triumph in another. Given our differing childhood experiences and our varied attachment styles, some people are more open to the joy of the unexpected and the magic of mistakes than others. Some people have little cognitive flexibility and are exceptionally slow to accept that their plans have gone awry, something new is happening and there might be unexpected value, learning or happiness in the new situation. There very often is.

These kinds of thoughts raced through my mind last year as I enjoyed a book of travel mishaps, *How Not to Travel the World: Adventures of a Disaster-Prone Backpacker.*[2] In 2011 the author, Lauren Juliff, quit her job, sold everything she owned and left her home to go travelling. Juliff spent the following years visiting amazing places, but also experiencing a series of disasters including being robbed in Tanzania, scammed in China and falling into a leech-infested rice paddy in Indonesia. Time after time, Juliff encountered unfortunate circumstances, unexpected occurrences and simple bad luck that was as statistically unlikely as it was (it seems) inevitable.

But perhaps the most interesting aspect of Juliff's book is the joy that is to be found in such events. Juliff writes on her website that her bad luck might well have been the best thing that ever happened to her.[3] Juliff is clearly right, based on her writing, but realising this is by no means easy, especially if your bad luck has long-lasting consequences. In other words, while serious bad luck is entirely negative, there is a type or level of mistake that adds to our lives and helps us to explore, grow and be happy. These kinds of mishaps are impossible to predict, measure or study in a research project, but they are vital parts of who we are as people – and, as Juliff demonstrates, many such errors occur when we travel to foreign places and try to adjust to unfamiliar surroundings.

In 1997, emboldened by my trip to Japan, I decided to go to Russia in search of further adventure. In the intervening years I had travelled a good deal in Ireland, the United Kingdom, across Europe and to the United States, but Japan had ignited in me a hunger for the exotic. I yearned to go further afield. Russia seemed perfect: moderately eastern, totally unknown to me and sufficiently distant that I was confident of some culture shock to jolt me out of my comfortable west-European complacency.

I travelled to St Petersburg with a friend. We booked with Intourist, the venerable Russian tour operator that was founded in 1929 and still operates today. We flew with Aeroflot and stayed in a grand city-centre hotel. The snow and ice had melted, but the city was still windy, wintery and cold. St Petersburg is an imposing place, uncompromising and austere, more European than Asian, but filled with mystery.

We visited the chilly Hermitage museum, the beguiling Marinsky Theatre and the magnificent Peter and Paul Fortress. We took a trip to the Grand Palace at Peterhof, a short distance outside the city. The Peterhof palace and gardens were commissioned by Peter the Great in the early 1700s to rival Versailles. When we visited, the palace was partly shut and some of the garden statutes were still covered in wooden boxes to protect them from the frost. These touches made our visit more memorable: we were virtually alone in an impossibly elegiac landscape, the memory of winter fresh in the air.

At the hotel, besuited waiters treated us like royalty, trying to sell us caviar as we entered and left the dining room. At the subway, people offered to buy dollars and, when we said we had none, offered to sell us some. The trip was going smoothly until, one evening, a waiter recommended that we visit a night spot that I will call 'Jupiter'.

It is still not entirely clear to me what happened at Jupiter, but our visit there was the point at which our trip either slipped off the rails or became more interesting, depending on your point of view. Let me explain.

To get to Jupiter, we stepped into a taxi outside the hotel and handed the driver a piece of paper with the club's address on it. After a suspiciously long drive, for which the driver charged us virtually nothing, we arrived at the club and – as I recall – went downstairs. This might have been a mistake. For the following two hours, we had grave difficulty figuring out what was going on around us. There were people talking to us continually in Russian, even though we clearly did not speak a word of the language. We could not find a bar in the club, just a series of rooms with people sitting around. Some were dancing. We wandered about, utterly confused by the place, the people, the sounds and whatever was going on. We could not find an exit.

Our first thought was that we would be robbed or worse, but, at one point, someone handed my wallet to me. It seems that I had dropped it earlier and this person found it and gave it back. Nothing was stolen. Now we were more confused than ever. And still the music throbbed, people talked to us incessantly and we drifted through room after room wondering where everyone had bought their drinks, why some people were sitting watching television and why there were so many clashing sources of sound.

After a couple of hours we found an exit and left the club. But now we were both confused *and* lost, because we had no idea where we were. It was 2 a.m. in St Petersburg in the mid-1990s and we were utterly, utterly lost, without the faintest clue where to go or how to get back to our hotel.

Russia is a lot more dangerous than Japan, so we were worried. We'd drifted around in the dark for an hour when a car pulled up beside us. The man indicated that he was a taxi driver and gestured for us to get in. After some discussion, we decided it was safer to get in rather than continue to drift in the dark. The man then drove for almost an hour before stopping the car abruptly, lighting a cigarette and opening a can of beer.

So this was it: death. We were about to be murdered in a back-street of St Petersburg.

The driver spoke no English. He saw our consternation and, in fairness, tried to help by offering us two beers. We declined. He laughed, spoke in Russian and pointed ahead, into the darkness. We got out of the car and looked where he indicated. We could just about see the Neva river. This was immediately reassuring: at least we were still in St Petersburg. But then we saw the reason for the delay: the bridge ahead of us was raised to let ships through. We could not cross the river until the bridge came down.

Terrified, exhausted and relieved at the same time, we sat back into the car and – bizarrely – fell asleep. The driver woke us gently when he arrived back to our hotel at 6 a.m. He accepted his fare, declined a tip, and drove off laughing, opening another can of beer as he vanished from sight. The hotel doorman greeted us with a smile. He asked if we had a good night and offered to sell us some caviar. Or dollars: he also had dollars.

## Bangalore: 'I always knew where you were'

Accidents and mishaps come in all shapes and sizes. The evening I spent lost in Tokyo was gentle and revelatory. By way of contrast, my adventures in St Petersburg were, at first sight, memorable for all the wrong reasons. Who on earth heads out into the night in 1990s Russia with no idea where they are going or how to get home? What kind of traveller does not know that the Neva bridges are drawn by night, making it virtually impossible to cross the river? And who in their right mind steps into a strange car at 2 a.m. just because the driver, between gulps of beer, claims that he drives a taxi?

But the night in St Petersburg was memorable for all the right reasons, too. This was another of the few occasions when all my moorings were lost and I was completely adrift in the world. Unlike the evening in Japan, my experience in St Petersburg was thrilling in a terrifying way, rather than a dreamy, dislocated way. But it was still a wonderful adventure, as a series of mistakes and misjudgements provided an experience that nourished my sense of self and contributed greatly to my happiness.

Today Russia is a very different place. I read that Jupiter, one of the oldest clubs in the city, was an underground legend in its heyday. It changed location since I visited and is now a music bar, café and venue for children's parties. I cannot for the life of me account for my disorientation the night that we visited: the club was lovely, if strange; there was no alcohol or drugs involved (I use neither); and my travelling companion was (and still is) a model of good sense, sobriety and wisdom.

But I would not change a moment of that night even if I could, because I remember it as a rare, valuable experience. On the face of it, there are many ways to get lost, so we should all be getting lost all the time. We can be lost because we don't know where we are, because we don't know where we're going or because we don't know what is happening around us. In St Petersburg we were all three kinds of lost at the same time. The travel, the night-time and the exoticism of Russia was an exhilarating cocktail that simply swept us away.

But despite all these ways of getting lost, we do not get lost nearly often enough. Is there a way to increase the chances of getting securely lost, so that we enjoy the strangeness of lostness without the fear of being killed by shifty taxi drivers?

For me, the first step is to seek out distant locations that will open me to the idea of adventure, lower my usual defences and then just see what happens. Remember, your adventure just needs to be an adventure by your standards, not by anyone else's standards. An incident that some people might find banal could be a huge adventure to you. What matters is how it feels to you: did you feel lost and adrift, with no idea what to do next? If so, did you feel totally secure (which is good), a little nervous (which is better) or so frightened that any possible happiness was knocked out of the experience (which is bad)? The right balance is difficult to achieve, but the balance is about *you* and not about anyone else.

I have one final tale about creating space for unmoored drifting though the world and the joys that it can bring. In 2016 I found myself in Bangalore in India. I select my travel destinations with care, to open myself to some adventure along the way. Bangalore did not disappoint, as I managed to get well and truly lost once again.

Bangalore is big. The city has a population of around ten million people and stretches as far as the eye can see. It teems with life: crowds of people, endless traffic, cows, dogs and heaven knows what else. From the moment I set foot in the city, I was entranced. I fell hopelessly in love with India in an instant, and subsequent trips only deepened my enjoyment of this infinite country. Bangalore, Delhi, Mumbai, Pune, Bihar – you name it, I love it.[4]

But it was busy, bustling Bangalore that stole my heart on that first day. The morning after I arrived, I went to the hotel reception to find a taxi to take me to the city centre for some sightseeing. The exceptionally enthusiastic group of receptionists told me this would cost the equivalent of €20 in Indian rupees. This seemed expensive but I was staying around 30 kilometres outside the city centre, so I said yes. I went out to meet the driver, a tall, friendly man who spoke some English. His name was Vivaan. Only then did I realise that I had engaged both the car and the driver for the day. It did not seem like a high price for all that and Vivaan was happy with his day's work. He launched us into the traffic with gusto.

One hour and several near-death experiences later, we arrived at a beautiful Hindu temple, which Vivaan and I visited together in near silence. At his request, I took a photograph of Vivaan with his phone, so that he could prove to his wife that he had actually visited a temple. Vivaan promised to take me to see his family later on. I asked if we could visit Bangalore's vast market district and he smiled: 'Always the markets!' Vivaan shook his head in amusement and drove to the Krishna Rajendra Market, also known as City Market.

As I stepped out of the taxi, Vivaan told me to explore the market and its neighbouring streets, which were filled with hundreds of tiny shops, street vendors and all kinds of other attractions. The market area stretches for many kilometres. My phone had no coverage and I was clearly going to get lost, so I asked where I would meet him later on. Vivaan smiled and said: 'Don't worry, I will find you.' With that, he vanished into the traffic. I was on my own.

If you have not visited India, it is difficult to convey the disorienting cacophony that hits you in a market, train station or other crowded area. The noise alone is enough to sweep you away: people talking, shouting and singing, traffic thundering past, music blaring from shops, dogs barking, public announcements from loudspeakers. Now, all alone in Bangalore's market district, I was surrounded by tens of thousands of people, all talking or trying to sell me things.

There was a warren of streets, an infinity of people and several dozen cows. As far as I could see, the streets had no names. The smells from the market were pungent and ripe: people, flowers, vegetables, spices, animals and various less pleasant odours rolled into one intoxicating mix that some people find vile but I find exciting.

I faced a decision. Would I simply trust Vivaan and drift into the market and the apparent chaos of the surrounding streets, without the faintest idea where I was on a map? Could I be confident that Vivaan would somehow magically find me in this crowd of some hundreds of thousands of people? Maybe millions? Or should I panic and try to find my way home at once? I had read about solitary travellers being mugged or kidnaped in India. Bangalore was neither Tokyo nor St Petersburg, but it felt as foreign as Tokyo and as overwhelming as St Petersburg. Suddenly, it was decision time again: was I securely lost or was I being irresponsible?

I went into the flower market to think it over and that is where my decision was made. I don't know if it was the fragrance of the flowers or the smell of raw sewage that overcame me, but I quickly slipped into a mindset similar to the one I experienced in Tokyo. I began to drift through the flower market with no thought about where I was going, what I would do when I got there or how I would get home. It was as if I was in a dream or had developed a fugue state in which I was distanced from my identity and usual self. I just wandered on with little thought about the past or the immediate future.

Clearly, some part of me decided to trust that Vivaan would find me or that things would simply work out in some other way. I could not phone Vivaan or another taxi. My maps were useless. I had no idea where I was. But I felt oddly secure in my lostness. I felt, rather than knew, that all would be well.

On this basis, every other thought vanished from my mind and I rambled through rows of flower stalls, hallways filled with garlands of glorious blooms, and strange corridors of small warehouses that seemed to function as shops, flower-preparation areas and family homes. Everywhere I received smiles and nods. Several times, small children gave me flowers and then ran to hide behind their giggling mothers. I was the only non-Indian person in sight.

After an hour or so looking at flowers, I meandered down streets lined with shops. The market district seemed to go on for ever. In the end, I wandered around eight or nine kilometres away from where I started, though tiny streets, courtyards and crowded laneways. I hadn't the faintest idea where I was. It began to get dark. After about three hours, I ended up bargaining furiously with a street vendor over a rather large statue of the Buddha for which he was charging €300. I decided it was worth €40, so I started by offering €20. The vendor spoke no English, but we handed a calculator back and forth with our respective positions.

Just as we concluded the transaction (I paid €50), Vivaan magically appeared at my side. I was flabbergasted. I had literally no idea how he located me in a city of ten million people when I had walked so far from my drop-off point, through labyrinthine streets and crowds of tens of thousands of people. I asked Vivaan how he found me – had he followed me?

'No,' he laughed, 'I went home. But you are the only foreigner in the district today. They phoned me.'

'So I was being watched all the time?'

'Nobody followed you,' he said carefully, 'but we all saw where you went.'

'*We*? Do you know these people?' I asked, pointing at the man who had just sold me the Buddha, and his colleagues in the tiny shop.

'Yes,' said Vivaan, gesturing at the throngs of people around him: 'I know all of them.'

'All ten million of them?'

'Yes,' he repeated, grinning. 'I always knew where you were.'

## Flow: finding ways to lose ourselves

To get securely lost, we need both a willingness to wander off the path and some level of security that we will be found. In Bangalore, Vivaan, despite his exuberant driving style, gave me confidence that he would reappear when needed. I have no idea how he inspired such trust in me, but he did. In psychological terms, my relationship with Vivaan might be described as 'secure attachment', in that I felt free to wander into the unknown without too much worry. I knew that all would be well in the end. Vivaan would rescue me.

Attachment theory was devised by psychiatrist and psychoanalyst John Bowlby, who wrote that children are best able to explore the world when they have the knowledge of a secure base (such as a parent or guardian) to return to in times of need.[5] This is known as secure attachment, as opposed to anxious, ambivalent or avoidant attachment. Reliable parents or guardians who consistently or almost always respond to their child's needs tend to create securely attached children. These children are certain that their care-givers will be responsive to their needs and reappear when needed. Once I trusted Vivaan the way that a child trusts a reliable parent, I opened up a space within which I could be securely lost. I could safely wander off into the market and let it absorb all of my attention. I knew that everything would somehow work out well in the end.

It helps that the examples of lostness I have given all took place in exotic locations in which it was relatively easy to become disoriented, lose myself in my surroundings and let go of conscious awareness. In Tokyo, St Petersburg and Bangalore, it did not take much for me to forget precisely where I was, what I was doing and how I would get home. As it turned out on all three occasions, these were important, enjoyable experiences that contributed greatly to my happiness.

In all three cities, I both wandered off and felt oddly secure most of the time. In Tokyo, I was reassured by the low crime rate, friendly people and general dreaminess of the place. In St Petersburg, I made unwise decisions to trust complete strangers (such as hopping into an unmarked car at 2 a.m.) in order to feel security in my lostness. No matter how foolish such actions appear in retrospect, they made me feel more secure at the time. Also, my friend was with me. And in Bangalore, Vivaan functioned as a sort of guardian whom I decided to trust and who rewarded that trust by finding me in the market and even taking me to visit his family. I tipped him well but did not provide school fees for his children, for which he asked several times. I valued his help, but I have my limits.

It is not, however, necessary to travel to distant lands to lose ourselves. I would journey to all four corners of the earth in search of strangeness, culture shock and lostness, but nobody has the time or money to take a long trip every time they need to let their moorings slip for a while. Happily, there are other ways to lose ourselves much closer to home, chiefly by becoming deeply absorbed in activities that we love – so absorbed that the rest of the world vanishes from consciousness and there is only the rhythm of our feet pounding the pavement as we run, the motion of our arms in the water as we swim, the click of knitting needles or the 'empty mind' of meditation.

This is called 'flow' in Western traditions and 'no self' in Buddhist philosophy. It is a deeply nourishing mental state that is highly conducive to happiness. We need more of it.

Hungarian-American psychologist Mihaly Csikszentmihalyi points out that we are at our happiest when we are in this state of flow, which occurs when we are so engrossed in something that nothing else matters.[6] We enter this state when we get 'in the zone' doing an activity that we enjoy, that requires concentration and at which we are skilled. If all of these elements are present, we cease to worry about anything else, we are utterly absorbed in the activity at hand, and we emerge at the end refreshed and happier than when we started out.

To help us enter this state, the level of challenge and our level of skill must be finely balanced. We need to feel some challenge, but we also need to believe that we are accomplishing something. The challenge must not be so great that our efforts seem futile, but it must not be so small that we get no reward or the activity does not seem worthwhile. If everything clicks into place, we become so deeply absorbed in the activity that the rest of the world simply falls away and, if everything goes well, our very sense of self can vanish. We exist only insofar as we are performing the task at hand. Nothing else registers.

Building flow into our lives can help us to find purpose and enjoyment in what we do, keep on learning and better manage the challenges that life sends our way.[7] One way to do this is to list activities in which we lose ourselves in a state of flow and list activities that do not absorb us and that we dread.[8] By focusing on the activities that bring us into a state of flow, we achieve greater absorption in the moment, deeper engagement with our activities and higher levels of satisfaction and happiness in our lives.

This state of flow is described in many cultures, perhaps most notably in Buddhism, which speaks of 'no self'. This teaching does not mean that there is literally *no* self – clearly, my 'self' got up this morning, drank a cup of tea and is typing on this computer.[9] No self means that the idea of 'self' feels more concrete and unchanging to us than it really is. Our 'selves' are constantly changing and only exist in the first place because of a fortunate coincidence of external circumstances and conditions, all of which could shift at any moment: we are excessively attached to the idea of 'self', which is not as permanent or unchanging as we commonly imagine. And thank goodness for that.

Entering a state of flow is one of the best ways to see no self in action. When I am utterly immersed in work on my laptop, I forget all kinds of important things: what time it is, whether or not I have fed the cat, when I need to leave for work, and what about that cup of tea I made half an hour ago, that now sits cold and untouched by the side of my keyboard? For other people, sport helps them reach this state of flow or no self: running, swimming, walking or exercising in a gym. For yet others, activities like meditation, yoga or even knitting can get them into the groove and let their troubles, worries and personal identities melt away, if only for a few hours.

This is possibly the most nourishing form of lostness: getting lost from ourselves in a healthy way. In general, we tend to overcontrol our lives. We struggle to let ourselves be lost. We hang on to the illusory certainties of everyday life. We fail to drift off as much as we should. Sometimes this is because we simply do not have the time to spend being lost, going running, sitting meditating or travelling to far-off places. But failing to get lost comes at considerable cost to our inner lives and happiness. We need to let go more, be it in the gym, at the yoga studio, deep in the markets of Bangalore or even here, at my laptop.

Getting lost is vital, if only so that we can be found again and return to our lives refreshed, with new ideas, new experiences and new ways of seeing our world.

## Top tips for happiness: losing yourself

- It is vital that we find a way to lose ourselves in a healthy way, if only so that we can find ourselves again or let ourselves be found by others.
- Getting 'securely lost' requires both a feeling of firm attachment that allows us to loosen the moorings in the first place and the ability to let go and drift away when the situation is right.
- Even if we do not travel to foreign climes, we can still lose ourselves in the moment if we become utterly absorbed in a given activity and enter a state of 'flow' or 'no self'. This is a deeply nourishing state of mind, highly conducive to happiness.
- We enter a state of flow when we choose a challenge, concentrate fully and use all our skills to perform the task at hand.
- The rest of the world and our 'selves' simply melt away, leaving just the motion of our legs as we run, the postures of our yoga practice, the movement of our arms as we swim or the sound of knitting needles.
- Prioritise these or similar activities in your life and spend as much time as possible pursuing them. This is the 'empty mind' of meditation. This is happiness. This is freedom.

# Conclusion

There are many ways to be happy. Albert Einstein said that 'a calm and modest life brings more happiness than the pursuit of success combined with constant restlessness'. This approach to well-being underpins much of the advice in this book, ranging from the findings of happiness research to the joys of being securely lost, from the importance of dreams to the value of politics, from the merits of physical exercise to the value of simply sitting in meditation, seeking our 'empty mind'.

Ultimately, there are any number of different perspectives on happiness, perhaps as many perspectives as there are people on earth.[1] Against this background, this book has focused chiefly (but not exclusively) on scientific evidence about happiness and suggested a series of measures to increase well-being in our lives. This final chapter summarises key steps that can help us to build a life that is more hospitable to happiness.

## Steps to happiness

Chapter 1 of this book began with an overview of research linking happiness with different aspects of who we are: our gender, age, genes, upbringing and geography. Women and men are just as happy (or unhappy) as each other; happiness is associated with childhood and later life; there is a midlife dip in our 40s; genes account for up to 50 per cent of the variance in happiness between individuals (we seem to have a genetic 'set point' for happiness); and childhood has a substantial effect on well-being: abuse, neglect and conflict in the home all corrode future happiness. Geography matters: Finland is the happiest country in the world, closely followed by many of its Nordic neighbours.

Chapter 2 focused on the relationships between happiness and certain aspects of what we do with our lives: starting a family, earning money, gaining or losing employment, looking after our physical health and holding particular religious beliefs or political positions. A birth boosts the parents' well-being for two years (up to a maximum of two births); well-being is generally lower among parents compared to non-parents; and children will not automatically make us happy, although they have the potential to boost our well-being in the right circumstances.

Earning money increases happiness, but the benefits diminish sharply beyond an annual income of around $95,000. How we perceive our income compared to the incomes of other people also matters, as do how we obtain and spend our money. Unemployment causes unhappiness. An increase in the national unemployment rate makes *everyone* unhappy, employed and unemployed alike. Happiness is associated with good physical and mental health (which overlap considerably), religion and spirituality (which also overlap) and holding political beliefs: conservatives consistently report themselves as happier than liberals.

After a grounding in the field of happiness research, we brought the research findings of Chapters 1 and 2 into the areas of well-being, psychology and spirituality. Based on these sources, Chapter 3 presented six overarching principles of a happy life: balance (seeking moderation in all things); love (loving both ourselves and others); acceptance (accepting what we cannot change; changing what we can); gratitude (we are lucky to be alive); avoiding comparisons with other people (which are the root of most human unhappiness); and believing in something that matters to us: politics, religion, philosophy, football or even the emotional lives of minor celebrities (if you absolutely must). It does not really matter what we believe, as long as we *believe.*

The next seven chapters took research findings and insights from the first three chapters and used them to inform practical strategies to increase happiness in our day-to-day lives. Here's what to do.

Chapter 4 started by looking at sleeping and waking. Without sleep, there can be no lasting happiness. Adults need between seven and nine hours' sleep in every 24 hours. We should keep our bedrooms dark, cool, comfortable and free of distractions; avoid naps during the day; and ease ourselves into sleep in the evening. We should avoid stimulants in the hours before bed (e.g. coffee, alcohol, sugar, cigarettes) and enrich our diets with foods that contain tryptophan (e.g. chicken, turkey, milk, dairy, nuts, seeds). For serious or persistent sleep problems, medical advice might be needed. Finally, just as we need to learn to sleep, we need to learn to wake up. This means waking up promptly, exercising (ideally outside in the morning) and 'awakening' more to the worlds around us and within us in our daily lives, as advised by virtually every philosophical and spiritual tradition on the planet. They are all correct. We need to wake up better and more.

Chapter 5 moved on to look at dreams, both the dreams we have at night while we sleep and the dreams we have during the day. Both matter. People who are prevented from dreaming at night develop psychological problems quite quickly. People who dream too little or too much during the day lose touch with reality in a different way. Despite their strangeness, dreams are full of meaning, often representing the fulfilment of wishes. We should reflect on key themes that recur, but not obsess over every detail. Daydreams matter just as much as night-dreams. Letting our minds wander helps us to think more creatively about our problems and imagine our futures more broadly. This is good for happiness, provided we chose our timing with care: daydreaming during a tedious meeting is perfectly acceptable; daydreaming while driving a car is not. Broadly, we do not dream enough. We need to rediscover the boldness to dream.

Chapter 6 looked at food. How can we eat intelligently so as to enhance our nutrition and boost our health and happiness? First, we need to de-link food from meanings, emotions and self-esteem, and see food for precisely what it is: just food. Second, we should follow official dietary guidance from trusted sources, such as Ireland's HSE, the United Kingdom's NHS and the United States' Centers for Disease Control and Prevention. There are certain nutrients, such as vitamin B12, that we tend to neglect, but that can help with brain health. We should make a special effort with these, but not obsess. A generally balanced diet will usually deliver all that our bodies need. Third, it is useful to harness the power of habit to improve our eating patterns. Moderate exercise goes hand in hand with sustainable dietary change, as our bodies simply demand the nutrients we need to keep going.

It is useful to reconnect with the sources of our food, try to grow some of our own supplies and note the seasonality of what we eat, possibly by gardening or cultivating an allotment. Finally, mindfulness can help with our diet, although it does not have all the answers. We should avoid distractions as we prepare and consume our food. When we eat, we should *just eat.* This is mindful eating. We need to follow our intuition, listen to our bodies and give thanks for our food and the ability to eat it. These are real gifts.

Chapter 7 focused on physical activity, on the basis that physical and mental health are intimately related to each other. Adults need to be physically active every day. We should do at least 150 minutes of moderate-intensity activity each week or 75 minutes of vigorous activity. We should also fit in some strengthening activities that work all our major muscles (legs, hips, back, abdomen, chest, arms and shoulders) two or three days each week.

Running is often presented as the ideal physical activity and, for some people, it is (although not me). Other possibilities include cycling, swimming, brisk walking, various other sports or even energetic forms of dance. The key to establishing a regular exercise habit is that our chosen activities should be sustainable, convenient, enjoyable and (ideally) sociable. Parkruns fulfil many of these requirements.[2] Finally, while physical activity is clearly conducive to well-being, it should be coupled with an ability to sit and rest. We need to both move and stop moving, consistent with Taoist ideas about achieving greater harmony and balance in our lives.

Chapter 8 looked at decluttering our lives and minds, which is just as important as decluttering our homes. We are busy people who need to slow down. This is difficult. Busyness is addictive. Our love of work has grown so out of proportion that we prefer restless activity to quiet stillness, pointless work to pleasant leisure. The problem is that we equate work with productivity, activity with value. We need to dismantle these beliefs.

Decluttering is the first step. We should identify the objects and activities that bring us joy and focus on them. This process of decluttering should be combined with an awareness of our motivations and the extent to which our current activities align with a sense of purpose in our lives. Intrinsic motivation works best. This means self-directing our activities when possible and prioritising activities that deliver inherent satisfaction and advance our well-being. It can be difficult to find the mental space to maintain a calm focus on what matters in the midst of our frenetic lives. Meditation helps. It takes time but is well worth the effort. Specific techniques such as mindfulness do not hold all the answers but are useful as part of a broader commitment to more contemplative, happier living.

Chapter 9 focused on connecting with, and disconnecting from, other people. A great deal of research shows that connecting with happy people creates happy networks and makes us happier. We need to both connect better with ourselves and feel part of something bigger, like leaves on trees. Disconnecting is equally important. Technology plays both positive and negative roles here, so we should manage it with care. We need to connect, disconnect and reconnect in a carefully managed way, especially in times of emergency or personal difficulty. With some thought, this is achievable and will help to make us happy.

Finally, Chapter 10 looked at the idea that it is regrettably difficult to get lost in the modern world, owing to the ubiquity of technology (especially smartphones) and a general excess of information. But getting 'securely lost' is important for our mental health, once we know we will be found. It is vital that we find a healthy way to lose ourselves, if only so that we can go through the process of finding ourselves again or let ourselves be found by others. I am hopelessly addicted to getting lost in exotic locations, but we can also lose ourselves by becoming deeply absorbed in activities we love: running, swimming, knitting or seeking the 'empty mind' of meditation. This is called 'flow' in western traditions and 'no self' in Buddhist philosophy. We all need more of this.

## A final word

Everyone wants to be happy, but the path to bliss is not always clear. Finding happiness often involves letting go of fixed ideas about what we think will make us happy and opening ourselves to new possibilities. In one sense, happiness cannot be purposely built. Happiness happens. But we can create the circumstances in which happiness is more likely to flourish in our lives and so increase our well-being.

There are many paths. Finding meaning in life is especially important.[3] I am fortunate that many aspects of my life provide meaning: spending time with my family and friends, doing clinical work with patients and their families, writing, researching and talking with the cat. It is worth noting that these are all *processes*, not *outcomes*: the activities that I value are as much about searching for meaning as actually finding it. Czech statesman and writer Václav Havel emphasised the journey rather than the destination: 'Keep the company of those who seek the truth; run from those who have found it.'

In psychiatry, agreeing on a path to meaning is a vital part of a person's recovery from mental illness or any state of profound psychological distress. What is it that matters to this person? Where do they find meaning? A person's valued goals are not always what you think they will be. Some people value friends and family vastly more than any meaning gleaned from education or work. For others, it is the reverse: they identify personal accomplishment as equally if not more important than other aspects of their lives. So be it.

Some other people simply value being able to leave the house or, at least, knowing that they are doing their best to do so, taking steps forward even if their ultimate goal remains distant. Again, it is effort, searching and direction that matter as much as reaching the destination. Process is as valuable as outcome, if not more so. Reaching a shared understanding about a person's therapeutic goals and ambitions lies at the heart of good mental health care that seeks to build a path to happiness in a person's life.

It is important to be patient and practical as we search for meaning and seek to increase our happiness. Positive thinking is important, but we also need to stay rooted in reality if we are to create lasting well-being.[4] The truth is our greatest strength, no matter how long it takes us to adjust to difficult realities. These things take time.

The various strategies outlined here combine research evidence with scientific, psychological and even spiritual advice in order to chart a happier path through our complex world. I hope you find it helpful.

It is not always possible to pursue happiness directly or explicitly. Sometimes, it is better to focus on creating the conditions for general well-being in our lives, in the confident belief that happiness will follow. It will. Entering a state of flow is especially helpful. This means becoming utterly absorbed in an activity or situation so that the rest of the world simply melts away and we enter a state of 'no self'. In the end, that is the final message of this book: we should seek out healthy ways to lose ourselves and the world, if only for a period of time.

Only then will we be happy. Only then will we be free.

Leabharlanna Poiblí Chathair Baile Átha Cliath
Dublin City Public Libraries

# Notes

## Introduction

1 European Social Survey (ESS). Source Questionnaire. London: European Social Survey, 2018. https://www.europeansocialsurvey.org/methodology/ess_methodology/source_questionnaire, accessed 7 May 2020.

2 Blanchflower, D.G. *Is Happiness U-shaped Everywhere? Age and Subjective Well-being in 132 Countries (NBER Working Paper No. 26641).* Cambridge, MA: National Bureau of Economic Research, 2020. https://www.nber.org/papers/w26641, accessed 30 May 2020.

## One: Who We Are

1 Stevenson, B. and Wolfers, J. The paradox of declining female happiness. *American Economic Journal: Economic Policy* 2009; 1: 190–225. See also: General Social Survey. *The General Social Survey.* Chicago, IL: NORC at the University of Chicago, 2020. https://gss.norc.org, accessed 7 May 2020; Leonhardt, D. He's happier, she's less so. *New York Times,* 26 September 2007. https://www.nytimes.com/2007/09/26/business/26leonhardt.html, accessed 30 May 2020.

2 Petherick, A. Gains in women's rights haven't made women happier. Why is that? *Guardian,* 18 Mary 2016. Available at: https://www.theguardian.com/lifeandstyle/2016/may/18/

womens-rights-happiness-well-being-gender-gap, accessed 6 May 2020. Courtesy of Guardian News & Media Ltd.

3 Krueger, A.B. Are we having more fun yet? Categorizing and evaluating changes in time allocation. *Brookings Papers on Economic Activity* 2007; 2: 193–217. https://www.brookings.edu/wp-content/uploads/2007/09/2007b_bpea_krueger.pdf, accessed 30 May 2020.

4 Doherty, A.M. and Kelly, B.D. Social and psychological correlates of happiness in 17 European countries. *Irish Journal of Psychological Medicine* 2010; 27: 130–4; European Social Survey (ESS). *ESS Round 5: European Social Survey Round 5 Data (2010). Data file edition 3.4.* NSD – Norwegian Centre for Research Data, Norway – Data Archive and distributor of ESS data for ESS ERIC. doi:10.21338/NSD-ESS5-2010. See also: Kelly, B.D. 'Happiness-deficit disorder'? Prevention is better than cure. *Psychiatric Bulletin* 2011; 35: 41-5. https://doi.org/10.1192/pb.bp.110.031500, accessed 9 August 2020.

5 European Social Survey (ESS). *ESS Round 9: European Social Survey Round 9 Data (2018). Data file edition 1.2.* NSD – Norwegian Centre for Research Data, Norway – Data Archive and distributor of ESS data for ESS ERIC. doi:10.21338/NSD-ESS9-2018; European Social Survey (ESS). *Source Questionnaire.* London: European Social Survey, 2018. https://www.europeansocialsurvey.org/methodology/ess_methodology/source_questionnaire, accessed 7 May 2020.

6 Qian, G. The effect of gender equality on happiness: Statistical modelling and analysis. *Health Care for Women International* 2017; 38: 75–90.

7 Blanchflower, D.G. *Is Happiness U-shaped Everywhere? Age and Subjective Well-being in 132 Countries (NBER Working Paper No. 26641).* Cambridge, MA: National Bureau of Economic Research, 2020. https://www.nber.org/papers/w26641, accessed 30 May 2020.

8 Rauch, J. *The Happiness Curve: Why Life Gets Better After Midlife.* London: Green Tree, 2018.

9 Myers, D.G. and Diener, E. Who is happy? *Psychological Science* 1995; 6: 10–19.

10 US National Library of Medicine. *What is a gene?* https://ghr.nlm.nih.gov/primer/basics/gene, accessed 6 May 2020.

11 Lykken, D. and Tellegen, A. Happiness is a stochastic phenomenon. *Psychological Science* 1996; 7: 186–9.

12 Okbay, A. et al. Genetic variants associated with subjective well-being, depressive symptoms, and neuroticism identified through genome-wide analyses. *Nature Genetics* 2016; 48: 624–33.13 Lyubomirsky, S. What influences our happiness the most? *Psychology Today*, 4 May 2008. https://www.psychologytoday.com/us/blog/how-happiness/200805/

what-influences-our-happiness-the-most, accessed 5 May 2020; Rowe, M. The basics of happiness. *Farmers Journal*, 7 March 2020.

14 Gaffney, M. *Flourishing: How to achieve a deeper sense of well-being, meaning and purpose – even when facing adversity.* Dublin: Penguin Ireland, 2011. See also: Haidt, J. *The Happiness Hypothesis: Putting Ancient Wisdom and Philosophy to the Test of Modern Science.* London: William Heinemann, 2006.

15 Layard, R. and Clark, D.M. *Thrive: The Power of Evidence-Based Psychological Therapies.* London: Allen Lane, 2014. See also: Layard, R. *Happiness: Lessons from a New Science.* London: Allen Lane, 2005; Layard, R. *Can We Be Happier?* London: Pelican, 2020.

16 Amato, P.R., Loomis, L.S. and Booth, A. Parental divorce, marital conflict, and offspring well-being during early adulthood. *Social Forces* 1995; 73: 895–915.

17 Jones, D.E., Greenberg, M. and Crowley, M. Early social-emotional functioning and public health: The relationship between kindergarten social competence and future wellness. *American Journal of Public Health* 2015; 105: 2283–90. https://ajph.aphapublications.org/doi/full/10.2105/AJPH.2015.302630, accessed 30 May 2020.

18 Fowler, J.H. and Christakis, N.A. Dynamic spread of happiness in a large social network: Longitudinal analysis over 20 years in the Framingham Heart Study. *BMJ* 2008; 337: a2338. https://www.bmj.com/content/337/bmj.a2338, accessed 30 May 2020.

19 Christakis, N. and Fowler. J. *Connected: The Amazing Power of Social Networks and How They Shape Our Lives.* London: HarperPress, 2010.

20 Seligman, M.E.P. *Flourish: A New Understanding of Happiness and Well-being – and How to Achieve Them.* London and Boston: Nicholas Brealy Publishing, 2011.

21 Russell, H. *The Atlas of Happiness.* London: Two Roads, 2018.

22 European Social Survey (ESS). *Source Questionnaire.* London: European Social Survey, 2018. https://www.europeansocialsurvey.org/methodology/ess_methodology/source_questionnaire, accessed 7 May 2020.

23 Gallup. *World Poll Methodology.* Washington, DC: Gallup, 2020. https://news.gallup.com/poll/105226/world-poll-methodology.aspx, accessed 7 May 2020.

24 Helliwell, J.F., Layard, R., Sachs J.D. and De Neve, J-E. (eds). *World Happiness Report 2020.* Paris: Sustainable Development Solutions Network, 2020. https://worldhappiness.report/ed/2020, accessed 7 May 2020.

25 Helliwell, J.F., Layard, R. and Sachs, J.D. (eds). *World Happiness Report 2019*. Paris: Sustainable Development Solutions Network, 2019. https://worldhappiness.report/ed/2019, accessed 7 May 2020.

26 This chapter of the report was written by Frank Martela, Bent Greve, Bo Rothstein and Juho Saari.

**Two: What We Do**

1 Twenge, J.M., Campbell, W.K. and Foster, C.A. Parenthood and marital satisfaction: A meta-analytic review. *Journal of Marriage and Family* 2003; 65: 574–83.

2 Wolfinger, N.H. *Does Having Children Make People Happier in the Long Run?* Charlottesville, VA: Institute for Family Studies, 2018. https://ifstudies.org/blog/does-having-children-make-people-happier-in-the-long-run, accessed 8 May 2020.

3 Kamiya, Y., Akpalu, B., Mahama, E., Ayipah, E.K., Owusu-Agyei, S., Hodgson, A., Shibanuma, A., Kikuchi, K., Jimba, M. and Ghana EMBRACE Implementation Research Project Team. The gender gap in relation to happiness and preferences in married couples after childbirth: Evidence from a field experiment in rural Ghana. *Journal of Health, Population and Nutrition* 2017; 36: 8. https://link.springer.com/article/10.1186/s41043-017-0084-2, accessed 30 May 2020.

4 Myrskylä, M. and Margolis, R. Happiness before and after the kids. *Demography* 2014; 51: 1843–66.

5 Glass, J., Simon, R.W. and Andersson, M.A. Parenthood and happiness: Effects of work-family reconciliation policies in 22 OECD countries. *American Journal of Sociology* 2016; 122: 886–929. https://www.ncbi.nlm.nih.gov/pmc/articles/PMC5222535, accessed 30 May 2020.

6 Becker, C., Kirchmaier, I. and Trautmann, S.T. Marriage, parenthood and social network: Subjective well-being and mental health in old age. *PLoS One* 2019; 14: e0218704. https://journals.plos.org/plosone/article?id=10.1371/journal.pone.0218704, accessed 30 May 2020. See also: Wolfinger, N.H. *Does Having Children Make People Happier in the Long Run?* Charlottesville, VA: Institute for Family Studies, 2018. https://ifstudies.org/blog/does-having-children-make-people-happier-in-the-long-run, accessed 8 May 2020.

7 Malik, K. Does having children make you happy? Yes, if you let them. *Observer*, 25 August 2019. https://www.theguardian.com/commentisfree/2019/aug/25/does-having-children-make-you-happy-yes-if-you-let-them, accessed 8 May 2020. (Courtesy of Guardian News & Media Ltd.)

8 Powdthavee, N. Think having children will make you happy? *The Psychologist* 2009; 22: 308–10. https://thepsychologist.bps.org.uk/volume-22/edition-4/think-having-children-will-make-you-happy, accessed 30 May 2020.
9 Layard, R. *Happiness: Lessons from a New Science.* London: Allen Lane, 2005.
10 Diener, E., Ng, W., Harter, J. and Arora, R. Wealth and happiness across the world: material prosperity predicts life evaluation, whereas psychosocial prosperity predicts positive feeling. *Journal of Personality and Social Psychology* 2010; 99: 52–61. https://pdfs.semanticscholar.org/40ee/a5e85514360e7efb7de36502937c5e16f20d.pdf, accessed 30 May 2020.
11 Kahneman, D. and Deaton, A. High income improves evaluation of life but not emotional well-being. *Proceedings of the National Academy of Sciences of the United States of America* 2010; 107: 16489–93. https://www.pnas.org/content/107/38/16489, accessed 30 May 2020.
12 Jebb, A.T., Tay, L., Diener, E. and Oishi, S. Happiness, income satiation, and turning points around the world. *Nature Human Behaviour* 2018; 2: 33–8.
13 Hussain, D. Conceptual referents, personality traits and income-happiness relationship: An empirical investigation. *Europe's Journal of Psychology* 2017; 13: 733–48. https://www.ncbi.nlm.nih.gov/pmc/articles/PMC5763460, accessed 30 May 2020.
14 Boyce, C.J., Brown, G.D.A. and Moore, S.C. Money and happiness: Rank of income, not income, affects life satisfaction. *Psychological Science* 2010; 21: 471–5. See also: Oishi, S., Kesebir, S. and Diener, E. Income inequality and happiness. *Psychological Science* 2011; 22: 1095–100; Yu, Z. and Wang, F. Income inequality and happiness: An inverted U-shaped curve. *Frontiers in Psychology* 2017; 8: 2052. https://www.frontiersin.org/articles/10.3389/fpsyg.2017.02052/full, accessed 30 May 2020; Kelley, J. and Evans, M.D.R. The new income inequality and well-being paradigm: Inequality has no effect on happiness in rich nations and normal times, varied effects in extraordinary circumstances, increases happiness in poor nations, and interacts with individuals' perceptions, attitudes, politics, and expectations for the future. *Social Science Research* 2017; 62: 39–74.
15 Oishi, S., Kushlev, K. and Schimmack, U. Progressive taxation, income inequality, and happiness. *American Psychologist* 2018; 73: 157–68.
16 Donnelly, G.E., Zheng, T., Haisley, E. and Norton, M.I. The amount and source of millionaires' wealth (moderately) predict their happiness. *Personality and Social Psychology Bulletin* 2018; 44: 684–99.

17 Dunn, E.W., Aknin, L.B. and Norton, M.I. Spending money on others promotes happiness. *Science* 2008; 319: 1687–8.

18 Easterlin, R.A., McVey, L.A., Switek, M., Sawangfa, O. and Zweig, J.S. The happiness-income paradox revisited. *Proceedings of the National Academy of Sciences of the United States of America* 2010; 107: 22463–8. https://www.pnas.org/content/107/52/22463, accessed 30 May 2020.

19 Oishi, S., Kesebir, S. and Diener, E. Income inequality and happiness. *Psychological Science* 2011; 22: 1095–100.

20 Hussain, D. Conceptual referents, personality traits and income-happiness relationship: An empirical investigation. *Europe's Journal of Psychology* 2017; 13: 733–48. https://www.ncbi.nlm.nih.gov/pmc/articles/PMC5763460/, accessed 30 May 2020.

21 Yu, Z. and Wang, F. Income inequality and happiness: An inverted U-shaped curve. *Frontiers in Psychology* 2017; 8: 2052.

22 Clark, A.E. *A Note on Unhappiness and Unemployment Duration (Discussion Paper No. 2406).* Bonn: Forschungsinstitut zur Zukunft der Arbeit/Institute for the Study of Labour, 2006. http://ftp.iza.org/dp2406.pdf, accessed 20 May 2020.

23 Lucas, R.E., Clark, A.E., Georgellis, Y. and Diener, E. Unemployment alters the set point for life satisfaction. *Psychological Science* 2004; 15: 8–13.

24 De Neve, J-E. and Ward, G. Does work make you happy? Evidence from the World Happiness Report. *Harvard Business Review*, 20 March 2017. https://hbr.org/2017/03/does-work-make-you-happy-evidence-from-the-world-happiness-report, accessed 20 May 2020.

25 Bangham, G. *Happy Now? Lessons for Economic Policy Makers from a Focus on Subjective Well-Being.* London: Resolution Foundation, 2019. https://www.resolutionfoundation.org/publications/happy-now-lessons-for-economic-policy-makers-from-a-focus-on-subjective-well-being, accessed 20 May 2020.

26 Baggini, J. The secret to happiness? Health, housing and job security. *Guardian*, 14 February 2019. https://www.theguardian.com/commentisfree/2019/feb/14/happiness-health-housing-job-security-uk-citizens, accessed 20 May 2020. (Courtesy of Guardian News & Media Ltd.) See also: Rawlinson, K. People are happiest at ages of 16 and 70 in UK, says study. *Guardian*, 13 February 2019. https://www.theguardian.com/global/2019/feb/13/people-are-happiest-at-ages-of-16-and-70-in-uk-says-study, accessed 20 May 2020. See also: Helliwell, J.F. and Aknin, L.B. Expanding the social science of happiness. *Nature Human Behaviour* 2018; 2: 248–52.

27 Marmot, M., Allen, J., Goldblatt, P, Boyce, T., McNeish, D., Grady, M. and Geddes, I. *Fair Society, Healthy Lives: The Marmot Review. Strategic Review of Health Inequalities in England post-2010*. London: The Marmot Review, 2010. http://www.instituteofhealthequity.org/resources-reports/fair-society-healthy-lives-the-marmot-review/fair-society-healthy-lives-full-report-pdf.pd, accessed 20 May 2020.

28 Tay, L. and Kuykendall, L. Promoting happiness: The malleability of individual and societal subjective well-being. *International Journal of Psychology* 2013; 48: 159–76.

29 Bakker, A.B. and Oerlemans, W.G.M. Momentary work happiness as a function of enduring burnout and work engagement. *Journal of Psychology* 2016; 150: 755–78. https://www.tandfonline.com/doi/full/10.1080/00223980.2016.1182888, accessed 30 May 2020.

30 Oerlemans, W.G.M. and Bakker, A.B. Motivating job characteristics and happiness at work: A multilevel perspective. *Journal of Applied Psychology* 2018; 103: 1230–41.

31 Weziak-Bialowolska, D., Bialowolski, P., Sacco, P.L., VanderWeele, T.J. and McNeely, E. Well-being in life and well-being at work: Which comes first? Evidence from a longitudinal study. *Frontiers in Public Health* 2020; 8: 103 https://www.frontiersin.org/articles/10.3389/fpubh.2020.00103/full, accessed 30 May 2020.

32 Layard, R. *Happiness: Lessons from a New Science*. London: Allen Lane, 2005.

33 Doherty, A.M. and Kelly, B.D. When Irish eyes are smiling: Income and happiness in Ireland, 2003–2009. *Irish Journal of Medical Science* 2013; 182: 113–9. See also: European Social Survey (ESS). *ESS Round 1: European Social Survey Round 1 Data (2002). Data file edition 6.6.* NSD – Norwegian Centre for Research Data, Norway – Data Archive and distributor of ESS data for ESS ERIC. doi:10.21338/NSD-ESS1-2002; European Social Survey (ESS). *ESS Round 2: European Social Survey Round 2 Data (2004). Data file edition 3.6.* NSD – Norwegian Centre for Research Data, Norway – Data Archive and distributor of ESS data for ESS ERIC. doi:10.21338/NSD-ESS2-2004; European Social Survey (ESS). *ESS Round 3: European Social Survey Round 3 Data (2006). Data file edition 3.7.* NSD – Norwegian Centre for Research Data, Norway – Data Archive and distributor of ESS data for ESS ERIC. doi:10.21338/NSD-ESS3-2006; European Social Survey (ESS). *ESS Round 4: European Social Survey Round 4 Data (2008). Data file edition 4.5.* NSD – Norwegian Centre for Research Data, Norway – Data Archive and distributor of ESS data for ESS ERIC. doi:10.21338/NSD-ESS4-2008.

34 Mwinnyaa, G., Porch, T., Bowie, J. and Thorpe, R.J., Jr. The association between happiness and self-rated physical health of African American men: A population-based cross-sectional study. *American Journal of Men's Health* 2018; 12: 1615–20. https://www.ncbi.nlm.nih.gov/pmc/articles/PMC6142117, accessed 30 May 2020.

35 Peiró, A. Happiness, satisfaction and socio-economic conditions: Some international evidence. *Journal of Socio-Economics* 2006; 35: 348–65.

36 Steptoe, A. Happiness and health. *Annual Review of Public Health* 2019; 40: 339–59. https://www.annualreviews.org/doi/pdf/10.1146/annurev-publhealth-040218-044150, accessed 30 May 2020.

37 Steptoe, A., Deaton, A. and Stone, A.A. Subjective well-being, health, and ageing. *Lancet* 2015; 385: 640–8. https://www.ncbi.nlm.nih.gov/pmc/articles/PMC4339610, accessed 30 May 2020. See also: Steptoe, A., Breeze, E., Banks, J., Nazroo, J. Cohort profile: The English Longitudinal Study of Ageing. *International Journal of Epidemiology* 2013; 42: 1640–8. https://www.ncbi.nlm.nih.gov/pmc/articles/PMC3900867, accessed 30 May 2020.

38 Siahpush, M., Spittal, M. and Singh, G.K. Happiness and life satisfaction prospectively predict self-rated health, physical health, and the presence of limiting, long-term health conditions. *American Journal of Health Promotion* 2008; 23: 18–26.

39 Diener, E., Pressman, S.D., Hunter, J. and Delgadillo-Chase, D. If, why, and when subjective well-being influences health, and future needed research. *Applied Psychology: Health and Well-Being* 2017; 9: 133-67. https://iaap-journals.onlinelibrary.wiley.com/doi/full/10.1111/aphw.12090, accessed 30 May 2020.

40 Park, N., Peterson, C., Szvarca, D., Vander Molen, R.J., Kim, E.S. and Collon, K. Positive psychology and physical health: Research and applications. *American Journal of Lifestyle Medicine* 2016; 10: 200–6. https://www.ncbi.nlm.nih.gov/pmc/articles/PMC6124958, accessed 30 May 2020.

41 Mujcic, R. and Oswald, A.J. Evolution of well-being and happiness after increases in consumption of fruit and vegetables. *American Journal of Public Health* 2016; 106: 1504–10. https://www.ncbi.nlm.nih.gov/pmc/articles/PMC4940663, accessed 30 May 2020.

42 Lathia, N., Sandstrom, G.M., Mascolo, C. and Rentfrow, P.J. Happier people live more active lives: Using smartphones to link happiness and physical activity. *PLoS One* 2017; 12: e0160589. https://journals.plos.org/plosone/article?id=10.1371/journal.pone.0160589, accessed 30 May 2020.

43 Koenig, H.G. Religion, spirituality, and health: The research and clinical implications. *International Scholarly Research Network: Psychiatry* 2012: 278730. https://www.hindawi.com/journals/isrn/2012/278730, accessed 30 May 2020.

44 Abdel-Khalek, A.M. Religiosity, health and happiness: Significant relations in adolescents from Qatar. *International Journal of Social Psychiatry* 2014; 60: 656–61.

45 Kamiya, Y. et al. The gender gap in relation to happiness and preferences in married couples after childbirth: Evidence from a field experiment in rural Ghana. *Journal of Health, Population and Nutrition* 2017; 36: 8. https://link.springer.com/article/10.1186/s41043-017-0084-2, accessed 30 May 2020.

46 Francis, L.J., Ok, Ü. and Robbins, M. Religion and happiness: A study among university students in Turkey. *Journal of Religion and Health* 2017; 56: 1335–47.

47 Rizvi, M.A.K. and Hossain, M.Z. Relationship between religious belief and happiness: A systematic literature review. *Journal of Religion and Health* 2017; 56: 1561–82. For further discussion, see also: Sims, A. *Is Faith Delusion? Why Religion is Good for Your Health*. London: Continuum, 2009.

48 Myers, D.G. and Diener, E. The scientific pursuit of happiness. *Perspectives on Psychological Science* 2018; 13: 218–25.

49 Napier, J.L. and Jost, J.T. Why are conservatives happier than liberals? *Psychological Science* 2008; 19: 565–72. See also: Pabayo, R., Kawachi, I. and Muennig, P. Political party affiliation, political ideology and mortality. *Journal of Epidemiology and Community Health* 2015; 69: 423–31.

50 Taylor, P., Funk, C. and Craighill, P. *Are We Happy Yet?* Washington, DC: Pew Research Center, 2006. https://www.pewsocialtrends.org/2006/02/13/are-we-happy-yet, accessed 23 May 2020.

51 Burton, C.M., Plaks, J.E. and Peterson, J.B. Why do conservatives report being happier than liberals? The contribution of neuroticism. *Journal of Social and Political Psychology* 2015; 3: 89–102. https://jspp.psychopen.eu/article/view/117/pdf, accessed 30 May 2020.

52 Newman, D.B., Schwarz, N., Graham, J. and Stone, A.A. Conservatives report greater meaning in life than liberals. *Social Psychological and Personality Science* 2019; 10: 494–503.

53 Schlenker, B.R., Chambers, J.R., Le, B.M. Conservatives are happier than liberals, but why? Political ideology, personality, and life satisfaction. *Journal of Research in Personality* 2012; 46: 127–46.

54 Bump, P. Old conservatives are objectively happier than you. *Atlantic*, 25 July 2013. https://www.theatlantic.com/national/archive/2013/07/old-conservatives-are-objectively-happier-you/312899, accessed 23 May 2020; Khazan, O. Why conservatives find life more meaningful than liberals. *Atlantic*, 26 July 2018. https://www.theatlantic.com/science/archive/2018/07/why-conservatives-find-life-more-meaningful-than-liberals/566105, accessed 23 May 2020; Brooks, A.C. Why conservatives are happier than liberals. *New York Times*, 7 July 2012. https://www.nytimes.com/2012/07/08/opinion/sunday/conservatives-are-happier-and-extremists-are-happiest-of-all.html, accessed 23 May 2020; Wilson, M. Conservatives say they're happier, but liberals act happier. Here's why. *New Zealand Listener*, 21 May 2019. https://www.noted.co.nz/health/health-psychology/happier-conservatives-liberals-why, accessed 23 May 2020.

55 Stavrova, O., Luhmann, M. Are conservatives happier than liberals? Not always and not everywhere. *Journal of Research in Personality* 2016; 63: 29–35.

### Three: Six Principles of a Happy Life

1 Tzu, L. (transl. William Scott Wilson). *Tao Te Ching: A New Translation*. Boston and London: Shambala, 2012.

2 Gilbert, B. Four ways that Taoism can help us be happier. *Elephant Journal*, 22 July 2014. https://www.elephantjournal.com/2014/07/4-ways-that-taoism-can-help-us-be-happier-brandon-gilbert, accessed 3 June 2020.

3 For an application of some of these principles, see: Heap, L. Fostering happiness through balance and integration: A Garmin case study. *American Journal of Health Promotion* 2019; 33: 1217–21. https://journals.sagepub.com/doi/full/10.1177/0890117119878277d, accessed 3 June 2020. See also: Yiping, S. Taoist view of happiness: Its impacts on modern life. *Journal of Zhejiang University (Humanities and Social Sciences)* 2011; 41: 72–8. http://www.zjujournals.com/soc/EN/Y2011/V41/I1/72, accessed 3 June 2020.

4 Lin, D. *The Tao of Happiness: Stories from Chuang Tzu for Your Spiritual Journey*. New York: Jeremy, P. Tarcher/Penguin, 2015. See also: Blumberg, A. Let the 'Tao Of Happiness' be your guide to a joyful, fulfilled life. *Huffpost*, 29 October 2015. https://www.huffpost.com/entry/tao-of-happiness_n_5631433be4b063179910d2c8, accessed 3 June 2020.

5 Yang, D. and Zhou, H. The comparison between Chinese and western well-being. *Open Journal of Social Sciences* 2017; 5: 181–8. https://www.scirp.org/journal/PaperInformation.aspx?paperID=80745, accessed 3 June 2020.

6 Zhang, G., Veenhoven, R. Ancient Chinese philosophical advice: Can it help us find happiness today? *Journal of Happiness Studies* 2008; 9: 425–43. https://link.springer.com/article/10.1007/s10902-006-9037-y, accessed 3 June 2020.
7 Cleary, T. *Taoist Meditation: Methods for Cultivating a Healthy Mind and Body.* Boston and London: Shambala, 2000.
8 Lieberman, M.D. *Social: Why Our Brains Are Wired to Connect.* Oxford: Oxford University Press, 2015. See also: Henig, R.M. Linked in. *New York Times*, 1 November 2013. https://www.nytimes.com/2013/11/03/books/review/social-by-matthew-d-lieberman.html, accessed 7 June 2020; Smith, E.E. Social connection makes a better brain. *Atlantic*, 29 October 2013. https://www.theatlantic.com/health/archive/2013/10/social-connection-makes-a-better-brain/280934, accessed 7 June 2020.
9 Esch, T. and Stefano, G.B. Love promotes health. *Neuroendocrinology Letters* 2005; 26: 264–7. http://www.nel.edu/userfiles/articlesnew/NEL260305A13.pdf, accessed 7 June 2020.
10 Stefano, G.B. and Esch, T. Love and stress. *Neuroendocrinology Letters* 2005; 26: 173–4. http://www.nel.edu/userfiles/articlesnew/NEL260305E02.pdf, accessed 7 June 2020.
11 Esch, T. and Stefano, G.B. The neurobiology of love. *Neuroendocrinology Letters* 2005; 26: 175–92. http://www.nel.edu/userfiles/articlesnew/NEL260305R01.pdf, accessed 7 June 2020.
12 Esch, T. and Stefano, G.B. The neurobiology of love. *Neuroendocrinology Letters* 2005; 26: 175–92. http://www.nel.edu/userfiles/articlesnew/NEL260305R01.pdf, accessed 7 June 2020.
13 Ostrow, R. Neuroscientist Matthew Lieberman says love really does hurt. *Australian*, 17 July 2015. https://www.scn.ucla.edu/pdf/Ostrow-71715-Australian.pdf, accessed 7 June 2020.
14 Poerio, G.L., Totterdell, P., Emerson, L-M. and Miles, E. Love is the triumph of the imagination: Daydreams about significant others are associated with increased happiness, love and connection. *Consciousness and Cognition* 2015; 33: 135–44. https://www.sciencedirect.com/science/article/pii/S1053810014002451?via%3Dihub, accessed 7 June 2020.
15 Hayes, S.C. Acceptance and commitment therapy: Towards a unified model of behavior change. *World Psychiatry* 2019; 18: 226–7. https://www.ncbi.nlm.nih.gov/pmc/articles/PMC6502411/pdf/WPS-18-226.pdf, accessed 5 June 2020.
16 Webster, M. Introduction to acceptance and commitment therapy. *Advances in Psychiatric Treatment* 2011; 17: 309–16. https://www.cambridge.org/core/services/aop-cambridge-core/content/view/

D67B44FDED4147CA35C1AA8FA9D3DDA5/S1355514600014231a.pdf/introduction_to_acceptance_and_commitment_therapy.pdf, accessed 6 June 2020.

17 Blackledge, J.T. and Hayes, S.C. Emotion regulation in acceptance and commitment therapy. *Journal of Clinical Psychology* 2001; 57: 243–55.

18 Brach, T. *Radical Acceptance: Embracing Your Life with the Heart of a Buddha*. New York: Bantam Books, 2003.

19 Kelly, B.D. *The Doctor Who Sat For a Year*. Dublin: Gill, 2019.

20 Cunha, L.F., Pellanda, L.C. and Reppold, C.T. Positive psychology and gratitude interventions: A randomized clinical trial. *Frontiers in Psychology* 2019; 10: 584. https://www.frontiersin.org/articles/10.3389/fpsyg.2019.00584/full, accessed 9 June 2020.

21 Davis, D.E., Choe, E., Meyers, J., Wade, N., Varjas, K., Gifford, A., Quinn, A., Hook, J.N., Van Tongeren, D.R., Griffin, B.J. and Worthington, E.L. Thankful for the little things: A meta-analysis of gratitude interventions. *Journal of Counselling Psychology* 2016; 63: 20–31.

22 Oishi, S. and Gilbert, E.A. Current and future directions in culture and happiness research. *Current Opinion in Psychology* 2016; 8: 54–8.

23 Caputo, A. The relationship between gratitude and loneliness: The potential benefits of gratitude for promoting social bonds. *Europe's Journal of Psychology* 2015; 11: 323–34. https://www.ncbi.nlm.nih.gov/pmc/articles/PMC4873114/, accessed 9 June 2020.

24 Stephenson, W. Do the dead outnumber the living? *BBC News*, 4 February 2012. https://www.bbc.com/news/magazine-16870579 accessed 9 June 2020.

25 Gaffney, M. *Flourishing: How to achieve a deeper sense of well-being, meaning and purpose – even when facing adversity*. Dublin: Penguin Ireland, 2011.

26 Buscaglia, L. *Born for Love: Reflections on Loving*. New York: Fawcett, 1992.

27 Festinger, L. A theory of social comparison processes. *Human Relations* 1954; 7: 117–40.

28 Baldwin, M. and Mussweiler, T. The culture of social comparison. *Proceedings of the National Academy of Sciences of the United States of America* 2018; 115: E9067–74. https://www.pnas.org/content/pnas/115/39/E9067.full.pdf, accessed 11 June 2020.

29 Easterlin, R.A. Explaining happiness. *Proceedings of the National Academy of Sciences of the United States of America* 2003; 100: 11176–83. https://www.pnas.org/content/pnas/100/19/11176.full.pdf, accessed 11 June 2020.

30 Martela, F., Greve, B., Rothstein, B. and Saari, J. The Nordic Exceptionalism: What Explains Why the Nordic Countries are Constantly Among the Happiest in the World. In: Helliwell, J.F., Layard, R., Sachs, J.D., De Neve, J-E. (eds). *World Happiness Report 2020* (pp. 129–46). Paris: Sustainable Development Solutions Network, 2020. https://worldhappiness.report/ed/2020, accessed 11 June 2020.
31 Kim, J., Hong, E.K., Choi, I. and Hicks, J.A. Companion versus comparison: Examining seeking social companionship or social comparison as characteristics that differentiate happy and unhappy people. *Personality and Social Psychology Bulletin* 2016; 42: 311–22.
32 Biali Haas, S. How to stop comparing yourself to others. *Psychology Today*, 5 March 2018. https://www.psychologytoday.com/intl/blog/prescriptions-life/201803/how-stop-comparing-yourself-others, accessed 11 June 2020.
33 Frankl, V. *Man's Search for Meaning*. New York: Simon and Schuster, 1959 (first published in German in 1946).
34 Smith, E.E. *The Power of Meaning: The True Route to Happiness*. London: Rider/Ebury Publishing, 2017.

**Four: Sleep/Wake Up**

1 Hirshkowitz, M. et al. National Sleep Foundation's sleep time duration recommendations: Methodology and results summary. *Sleep Health* 2015; 1: 40-3. https://www.sleephealthjournal.org/action/showPdf?pii=S2352-7218%2815%2900015-7, accessed 19 June 2020; Hirshkowitz, M. et al. National Sleep Foundation's updated sleep duration recommendations: Final report. *Sleep Health* 2015; 1: 233-43; National Sleep Foundation. *How Much Sleep Do We Really Need?* Arlington, VA: National Sleep Foundation, 2020. https://www.sleepfoundation.org/articles/how-much-sleep-do-we-really-need, accessed 19 June 2020.
2 Walker, M. *Why We Sleep: The New Science of Sleep and Dreams*. London: Penguin, 2017.
3 Vandekerckhove, M. and Cluydts, R. The emotional brain and sleep: An intimate relationship. *Sleep Medicine Reviews* 2010; 14: 219–26.
4 Palmer, C.A. and Alfano, C.A. Sleep and emotion regulation: An organizing, integrative review. *Sleep Medicine Reviews* 2017; 31: 6–16.
5 Tempesta, D., Socci, V., De Gennaro, L. and Ferrara, M. Sleep and emotional processing. *Sleep Medicine Reviews* 2018; 40: 183–95.
6 King, V. *10 Keys to Happier Living: A Practical Handbook for Happiness*. London: Headline, 2016.

7 Ong, A.D., Kim, S., Young, S. and Steptoe, A. Positive affect and sleep: A systematic review. *Sleep Medicine Reviews* 2017; 35: 21–32.
8 Short, M.A., Booth, S.A., Omar, O., Ostlundh, L. and Arora, T. The relationship between sleep duration and mood in adolescents: A systematic review and meta-analysis. *Sleep Medicine Reviews* 2020; 52: 101311.
9 World Health Organization. *The ICD-10 Classification of Mental and Behavioural Disorders*. Geneva: World Health Organization, 1992; Kelly, B.D. *Mental Health in Ireland: The Complete Guide for Patients, Families, Health Care Professionals and Everyone Who Wants to Be Well*. Dublin: Liffey Press, 2017.
10 National Health Service. *Why Lack of Sleep is Bad for your Health*. London: National Health Service, 2018. https://www.nhs.uk/live-well/sleep-and-tiredness/why-lack-of-sleep-is-bad-for-your-health/, accessed 20 June 2020.
11 Burnett, D. *Happy Brain: Where Happiness Comes From, and Why*. New York and London: W. W. Norton and Company, 2018.
12 Lyubomirsky, S. *The How of Happiness: A Practical Guide to Getting the Life You Want*. London: Sphere, 2007.
13 Seldon, A. *Beyond Happiness: How to Find Lasting Meaning and Joy in All That You Have*. London: Yellow Kite, 2015.
14 Sleep Council. *Sleep Hygiene*. Skipton: Sleep Council, 2020. https://sleepcouncil.org.uk/advice-support/sleep-advice/sleep-hygiene, accessed 19 June 2020.
15 Kelly, B.D. Are you dreaming of a good night's sleep? *Irish Independent*, 5 February 2018. https://www.independent.ie/life/health-wellbeing/health-features/are-you-suffering-from-sleep-debt-six-simple-ways-to-make-sure-you-get-quality-shut-eye-36560043.html, accessed 19 June 2020.
16 English, J. Saying no to sedatives – a quality improvement approach. *Forum* 2020; 4: 12–14.
17 Kerouac, J. *Wake Up: A Life of the Buddha*. London: Penguin, 2008.
18 Harris, S. *Waking Up: Searching for Spirituality Without Religion*. London: Transworld, 2014.
19 Kabat-Zinn, J. *Full Catastrophe Living: Using the Wisdom of Your Body and Mind to Face Stress, Pain, and Illness* (revised and updated edition). New York: Bantam Books, 2013.

**Five: Dream/Stop Dreaming**

1 Solms, M. Neurobiology and the neurological basis of dreaming. *Handbook of Clinical Neurology* 2011; 98: 519–44.

2 Walker, M. *Why We Sleep: The New Science of Sleep and Dreams*. London: Penguin, 2017.

3 Casey, P. and Kelly, B.D. *Fish's Clinical Psychopathology: Signs and Symptoms in Psychiatry* (fourth edition). Cambridge: Cambridge University Press, 2019.

4 Manger, P.R. and Siegel, J.M. Do all mammals dream? *Journal of Comparative Neurology* 2020; 528: 3197–204.

5 Zwaka, H., Bartels, R., Gora, J., Franck, V., Culo, A., Götsch, M. and Menzel, R. Context odor presentation during sleep enhances memory in honeybees. *Current Biology* 2015; 25: 2869–74. https://doi.org/10.1016/j.cub.2015.09.069, accessed 27 June 2020; see also: Melnattur, K., Dissel, S. and Shaw, P.J. Learning and memory: Do bees dream? *Current Biology* 2015; 25: R1040–1. https://doi.org/10.1016/j.cub.2015.09.001, accessed 27 June 2020.

6 Freud, S. *The Interpretation of Dreams*. Leipzig and Vienna: Franz Deuticke, 1900. See also: Kelly, B.D. The Freud Project, year four: The Interpretation of Dreams. *Irish Medical Times*, 14 March 2014; Kelly, B.D. The Freud Project: On Dreams. *Irish Medical Times*, 20 February 2015.

7 Eichenlaub, J-B., van Rijn, E., Gaskell, M.G., Lewis, P.A., Maby, E., Malinowski, J.E., Walker, M.P., Boy, F. and Blagrove, M. Incorporation of recent waking-life experiences in dreams correlates with frontal theta activity in REM sleep. *Social Cognitive and Affective Neuroscience* 2018; 13: 637–47. https://academic.oup.com/scan/article/13/6/637/5032636, accessed 30 June 2020.

8 Hooper, R. We've started to uncover the true purpose of dreams. *New Scientist*, 18 July 2018. https://www.newscientist.com/article/mg23931873-300-weve-started-to-uncover-the-true-purpose-of-dreams, accessed 30 June 2020.

9 Breus, M.J. Why we dream what we dream. *Psychology Today*, 20 January 2015. https://www.psychologytoday.com/ie/blog/sleep-newzzz/201501/why-we-dream-what-we-dream, accessed 30 June 2020.

10 Killingsworth, M.A. and Gilbert, D.T. A wandering mind is an unhappy mind. *Science* 2010; 330: 932 https://science.sciencemag.org/content/330/6006/932.abstract, accessed 2 July 2020.

11 Mooneyham, B.W. and Schooler, J.W. The costs and benefits of mind-wandering: A review. *Canadian Journal of Experimental Psychology/Revue Canadienne de Psychologie Expérimentale* 2013; 67: 11–18.

12 Christoff, K., Gordon, A.M., Smallwood, J., Smith, R. and Schooler, J.W. Experience sampling during fMRI reveals default network and executive system contributions to mind-wandering. *Proceedings of the National Academy of Sciences of the United States of America* 2009; 106: 8719–24. https://www.pnas.org/content/pnas/early/2009/05/11/0900234106.full.pdf, accessed 2 July 2020. See also: University of British Columbia. Brain's problem-solving function at work when we daydream. *ScienceDaily*, 12 May 2009. www.sciencedaily.com/releases/2009/05/090511180702.htm, accessed 2 July 2020.

13 Debrot, M.T. Daydreams: Concealed and revealed in analysis. *Psychoanalytic Review* 2019; 106: 559–2. https://guilfordjournals.com/doi/pdf/10.1521/prev.2019.106.6.559, accessed 24 June 2020.

14 National Sleep Foundation. *How Much Sleep Do We Really Need?* Arlington, VA: National Sleep Foundation, 2020. https://www.sleepfoundation.org/articles/how-much-sleep-do-we-really-need, accessed 3 July 2020.

15 Cleantis, T. Freudian-express: Dreams, the royal road to the unconscious. *Psychology Today*, 14 January 2011. https://www.psychologytoday.com/ie/blog/freudian-sip/201101/freudian-express-dreams-the-royal-road-the-unconscious, accessed 3 July 2020.

16 Kerouac, J. *Wake Up: A Life of the Buddha*. London: Penguin, 2008.

17 Harris, S. *Waking Up: Searching for Spirituality Without Religion*. London: Transworld, 2014.

18 Freud S. *Five Lectures on Psycho-Analysis*. Leipzig and Vienna: Franz Deuticke, 1910.

## Six: Eat/Stop Eating

1 Kabat-Zinn, J. *Full Catastrophe Living: Using the Wisdom of Your Body and Mind to Face Stress, Pain, and Illness* (revised and updated edition). New York: Bantam Books, 2013.

2 World Health Organization. *Obesity*. Geneva: World Health Organization, 2020. https://www.who.int/topics/obesity/en, accessed 14 July 2020.

3 Robertson, S., Davies, M. and Winefield, H. Why weight for happiness? Correlates of BMI and SWB in Australia. *Obesity Research & Clinical Practice* 2015; 9: 609–12.

4 Godoy-Izquierdo, D., González-Hernández, J., Rodríguez-Tadeo, A., Lara, R., Ogallar, A., Navarrón, E., Ramírez, M.J., López-Mora, C. and Arbinaga F. Body satisfaction, weight stigma, positivity, and happiness among Spanish adults with overweight and obesity. *International Journal*

*of Environmental Research and Public Health* 2020; 17: E4186. https://doi.org/10.3390/ijerph17124186, accessed 14 July 2020.

5 World Health Organization. *The ICD-10 Classification of Mental and Behavioural Disorders.* Geneva: World Health Organization, 1992.

6 Treasure J. Eating Disorders. In: Cooper, C.L., Field, J., Goswami, U., Jenkins, R. and Sahakian, B.J. (eds), *Mental Capital and Wellbeing.* Chichester: Wiley-Blackwell, 2010 (pp. 893–904).

7 Kelly, B.D. *Mental Health in Ireland: The Complete Guide for Patients, Families, Health Care Professionals and Everyone Who Wants to Be Well.* Dublin: Liffey Press, 2017

8 Valente, J. *How to Live: What the Rule of St. Benedict Teaches Us about Happiness, Meaning and Community.* London: William Collins, 2018.

9 Health Service Executive. *How To Eat Well.* Dublin: Health Service Executive, 2019. https://www2.hse.ie/wellbeing/how-to-eat-well.html, accessed 16 July 2020.

10 National Health Service. *The Eatwell Guide.* London: National Health Service, 2019. https://www.nhs.uk/live-well/eat-well/the-eatwell-guide, accessed 16 July 2020.

11 Centers for Disease Control and Prevention. *Healthy Eating for a Healthy Weight.* Atlanta, GA: Centers for Disease Control and Prevention, 2020. https://www.cdc.gov/healthyweight/healthy_eating/index.html, accessed 14 July 2020.

12 Graham, T. and Ramsey, D. *The Happiness Diet: A Nutritional Prescription for a Sharp Brain, Balanced Mood, and Lean, Energized Body.* New York: Rodale Inc., 2011.

13 King, V. *10 Keys to Happier Living: A Practical Handbook for Happiness.* London: Headline, 2016.

14 Klein, S. *The Science of Happiness: How Our Brains Make Us Happy – and What We Can Do to Get Happier.* Melbourne and London: Scribe, 2006.

15 Harrison, C. *Anti-Diet: Reclaim Your Time, Money, Well-Being and Happiness Through Intuitive Eating.* London: Yellow Kite, 2019.

16 Hardman, I. *The Natural Health Service: What the Great Outdoors Can Do for Your Mind.* London: Atlantic Books, 2020.

17 Stuart-Smith, S. *The Well Gardened Mind: Rediscovering Nature in the Modern World.* London: William Collins, 2020.

18 Smith, D. *Grow Your Own Happiness: How to Harness the Science of Wellbeing for Life.* London: Aster, 2019.

19 King, V. *10 Keys to Happier Living: A Practical Handbook for Happiness.* London: Headline, 2016.

20 Seldon, A. *Beyond Happiness: How to Find Lasting Meaning and Joy in All That You Have*. London: Yellow Kite, 2015.
21 Orbach, S. *On Eating*. London: Penguin, 2002.

**Seven: Move/Stop Moving**

1 Lyubomirsky, S. *The How of Happiness: A Practical Guide to Getting the Life You Want*. London: Sphere, 2007.
2 Blumenthal, J.A. et al. Effects of exercise training on older patients with major depression. *Archives of Internal Medicine* 1999; 159: 2349–56. https://doi.org/10.1001/archinte.159.19.2349, accessed 8 August 2020.
3 Khazaee-Pool, M., Sadeghi, R., Majlessi, F. and Rahimi Foroushani, A. Effects of physical exercise programme on happiness among older people. *Journal of Psychiatric and Mental Health Nursing* 2015; 22: 47-57. https://doi.org/10.1111/jpm.12168, accessed 8 August 2020.
4 Xu, W., Sun, H., Zhu, B., Bai, W., Yu, X., Duan, R., Kou, C. and Li, W. Analysis of factors affecting the high subjective well-being of Chinese residents based on the 2014 China Family Panel Study. *International Journal of Environmental Research and Public Health* 2019; 16: 2566. https://doi.org/10.3390/ijerph16142566, accessed 8 August 2020.
5 De Bruin, E.I., Valentin, S., Baartmans, J.M.D., Blok, M. and Bögels, S.M. Mindful2Work the next steps: Effectiveness of a program combining physical exercise, yoga and mindfulness, adding a wait-list period, measurements up to one year later and qualitative interviews. *Complementary Therapies in Clinical Practice* 2020; 39: 101137. https://doi.org/10.1016/j.ctcp.2020.101137, accessed 8 August 2020.
6 Stuart-Smith, S. *The Well Gardened Mind: Rediscovering Nature in the Modern World*. London: William Collins, 2020.
7 Seligman, M.E.P. *Flourish: A New Understanding of Happiness and Well-being – and How to Achieve Them*. London and Boston: Nicholas Brealy Publishing, 2011.
8 Sui, X., LaMonte, M.J., Laditka, J.N., Hardin, J.W., Chase, N., Hooker, S.P. and Blair, S.N. Cardiorespiratory fitness and adiposity as mortality predictors in older adults. *JAMA* 2007; 298: 2507–16. https://doi.org/10.1001/jama.298.21.2507, accessed 8 August 2020.
9 Farren, C. *The U-Turn: A Guide to Happiness*. Dublin: Orpen Press, 2013.
10 Bok, D. *The Politics of Happiness: What Government Can Learn from the New Research on Well-being*. Princeton and Oxford: Princeton University Press, 2010.
11 Seldon, A. *Beyond Happiness: How to Find Lasting Meaning and Joy in All That You Have*. London: Yellow Kite, 2015.

12 Health Service Executive. *Physical Activity Guidelines.* Dublin: Health Service Executive. 2020. https://www.hse.ie/eng/about/who/healthwellbeing/our-priority-programmes/heal/physical-activity-guidelines, accessed 8 August 2020.

13 National Health Service. *Exercise.* London: National Health Service, 2019. https://www.nhs.uk/live-well/exercise, accessed 8 August 2020.

14 National Health Service. *Physical Activity Guidelines for Children (Under 5 Years).* London: National Health Service, 2019. https://www.nhs.uk/live-well/exercise/physical-activity-guidelines-children-under-five-years, accessed 8 August 2020; National Health Service. *Physical Activity Guidelines for Children and Young People.* London: National Health Service, 2019. https://www.nhs.uk/live-well/exercise/physical-activity-guidelines-children-and-young-people, accessed 8 August 2020; National Health Service. *Physical Activity Guidelines for Older Adults.* London: National Health Service, 2019. https://www.nhs.uk/live-well/exercise/physical-activity-guidelines-older-adults, accessed 8 August 2020.

15 Centers for Disease Control and Prevention. *Physical Activity.* Atlanta, GA: Centers for Disease Control and Prevention, 2020. https://www.cdc.gov/physicalactivity/index.html, accessed 8 August 2020.

16 Rowe, M. *A Prescription for Happiness: The Ten Commitments to a Happier, Healthier Life.* Waterford: Dr Mark Rowe, 2015.

17 National Economic and Social Council. *Well-being Matters: A Social Report for Ireland.* Dublin: National Economic and Social Council, 2009. http://files.nesc.ie/nesc_reports/en/NESC_119_vol_II_2009.pdf, accessed 25 October 2020.

18 King, V. *10 Keys to Happier Living: A Practical Handbook for Happiness.* London: Headline Publishing Group, 2016.

19 Kabat-Zinn, J. *Full Catastrophe Living: Using the Wisdom of Your Body and Mind to Face Stress, Pain, and Illness* (revised and updated edition). New York: Bantam Books, 2013.

20 Hardman, I. *The Natural Health Service: What the Great Outdoors Can Do for Your Mind.* London: Atlantic Books, 2020.

21 Klein, S. *The Science of Happiness: How Our Brains Make Us Happy – and What We Can Do to Get Happier.* Melbourne and London: Scribe, 2006.

22 See www.parkrun.com.

23 Stevinson, C. and Hickson, M. Changes in physical activity, weight and wellbeing outcomes among attendees of a weekly mass participation event: A prospective 12-month study. *Journal of Public Health* 2019; 41: 807–14. https://doi.org/10.1093/pubmed/fdy178, accessed 8 August 2020.

24 Chen, Y-C., Chen, C., Martínez, R.M., Etnier, J.L. and Cheng, Y. Habitual physical activity mediates the acute exercise-induced modulation of anxiety-related amygdala functional connectivity. *Scientific Reports* 2019; 9: 19787. https://doi.org/10.1038/s41598-019-56226-z, accessed 8 August 2020.

25 Gaffney, M. *Flourishing: How to achieve a deeper sense of well-being, meaning and purpose – even when facing adversity.* Dublin: Penguin Ireland, 2011

26 Harvard Medical School. *Harvard Health Letter: The Top 5 Benefits of Cycling.* Boston, MA: Harvard Health Publishing, 2020. https://www.health.harvard.edu/staying-healthy/the-top-5-benefits-of-cycling, accessed 8 August 2020.

27 Cavendish, C. Make time to rest or the world will steal the moment from you. *Financial Times*, 29/30 August 2020.

**Eight: Do/Stop Doing**

1 Aird, P. Are we too busy to be happy? *British Journal of General Practice* 2015; 65: 256. https://doi.org/10.3399/bjgp15X684961, accessed 12 August 2020.

2 O'Keane, O. *Ten Times Happier: How to Let Go of What's Holding You Back.* London: HQ, 2020.

3 Tartakovsky, M. Busyness: The new status symbol. *Psych Central*, 5 August 2017 https://psychcentral.com/lib/busyness-the-new-status-symbol, accessed 12 August 2020.

4 Brûlé, T. Editor's letter: Calm after the storm. *Monocle*, February 2020.

5 Kondo, M. *The Life-Changing Magic of Tidying: A Simple, Effective Way to Banish Clutter Forever.* London: Vermilion, 2014; Kondo, M. *Spark Joy.* London: Vermilion, 2016.

6 See https://konmari.com.

7 Kondo, M. and Sonenshein, S. *Joy at Work: Organizing Your Professional Life.* London: Bluebird Books for Life, 2020.

8 Lathia, N, Sandstrom, G.M., Mascolo, C. and Rentfrow, P.J. Happier people live more active lives: Using smartphones to link happiness and physical activity. *PLoS One* 2017; 12: e0160589. https://doi.org/10.1371/journal.pone.0160589, accessed 13 August 2020.

9 Reynolds, S. *Organised: Simple Ways to Declutter Your House, Your Schedule and Your Mind.* Dublin: Gill Books, 2017.

10 Hanson, R. *Hardwiring Happiness: How to Reshape Your Brain and Your Life.* London: Rider, 2013.

11 Pink, D.H. *Drive: The Surprising Truth About What Motivates Us.* New York: Riverhead Books, 2009. See also: Kelly, J.D. Your best life: What motivates you? *Clinical Orthopaedics and Related Research* 2019; 477: 509-11. https://doi.org/10.1097/CORR.0000000000000656, accessed 14 August 2020.

12 Stiglitz, J.E., Sen A. and Fitoussi, J-P. *Report by the Commission on the Measurement of Economic Performance and Social Progress.* Paris: Commission on the Measurement of Economic Performance and Social Progress, 2009.

13 Hall, B. Sarkozy strives for measure of happiness. *Financial Times,* 15 September 2009; *Economist.* Measuring What Matters. *Economist,* 19 September 2009; *Irish Times.* Happiness index (editorial). *Irish Times,* 16 September 2009.

14 Wiking, M. Q&A. *Monocle,* May 2020.

15 Kelly, B.D. *The Doctor Who Sat For a Year.* Dublin: Gill, 2019.

16 Kelly, B.D. *Mental Health in Ireland: The Complete Guide for Patients, Families,Health Care Professionals and Everyone Who Wants to Be Well.* Dublin: Liffey Press, 2017.

17 Creswell, J.D. Mindfulness interventions. *Annual Review of Psychology* 2017; 68: 491–516. See also: Dalai Lama, Cutler, H.C. *The Art of Happiness: A Handbook for Living.* London: Hodder and Stoughton, 1998; Ricard, M. *Happiness: A Guide to Developing Life's Most Important Skill.* New York: Little, Brown and Company, 2006; Loizzo, J. *Sustainable Happiness: The Mind Science of Well-Being, Altruism, and Inspiration.* New York and London: Routledge, 2012; Tang, Y-Y., Hölzel, B.K., Posner, M.I. The neuroscience of mindfulness meditation. *Nature Reviews: Neuroscience* 2015; 16: 213–25.

18 Purser, R.E. *McMindfulness: How Mindfulness Became the New Capitalist Spirituality.* London: Repeater Books, 2019; Sykes, P. *How Do We Know We're Doing It Right? Essays on Modern Life.* London: Hutchinson, 2020.

**Nine: Connect/Disconnect**

1 Russell, B. *The Conquest of Happiness.* London: George Allen & Unwin, 1930.

2 Smith, D. *Grow Your Own Happiness: How to Harness the Science of Wellbeing for Life.* London: Aster, 2019.

3 Barry, H. *Emotional Healing: How To Put Yourself Back Together Again.* London: Orion Spring, 2020.

4 Fowler, J.H. and Christakis, N.A. Dynamic spread of happiness in a large social network: Longitudinal analysis over 20 years in the Framingham

Heart Study. *BMJ* 2008; 337: a2338. https://www.bmj.com/content/337/bmj.a2338, accessed 20 August 2020. See also: Tedstone Doherty, D., Moran, R., Kartalova-O'Doherty, Y. and Walsh, D. *HRB National Psychological Wellbeing and Distress Survey: Baseline Results.* Dublin: Health Research Board, 2007.

5 Christakis, N. and Fowler, J. *Connected: The Amazing Power of Social Networks and How They Shape Our Lives.* London: Harper Press, 2010.

6 Lieberman, M.D. *Social: Why Our Brains Are Wired to Connect.* Oxford: Oxford University Press, 2015.

7 Barry, H. *Self-Acceptance: How to Banish the Self-Esteem Myth, Accept Yourself Unconditionally and Revolutionise Your Mental Health.* London: Orion Spring, 2019.

8 Smith, D. *Grow Your Own Happiness: How to Harness the Science of Wellbeing for Life.* London: Aster, 2019.

9 Lyubomirsky, S. *The Myths of Happiness: What Should Make You Happy, but Doesn't, What Shouldn't Make You Happy, but Does.* New York: Penguin Press, 2013.

10 Brandreth, G. How to be happy. *Sunday Telegraph*, 30 January 2000; Brandreth, G. *Something Sensational to Read in the Train: The Diary of a Lifetime.* London: John Murray, 2009; Brandreth, G. *The 7 Secrets of Happiness: An Optimist's Journey.* London: Short Books, 2013.

11 Kelly, B.D. and Houston, M. *Psychiatrist in the Chair: The Official Biography of Anthony Clare.* Dublin: Merrion Press, 2020.

12 MacLachlan, M. and Hand, K. *Happy Nation? Prospects for Psychological Prosperity in Ireland.* Dublin: The Liffey Press, 2013.

13 Twenge, J.M., Martin, G.N. and Campbell, W.K. Decreases in psychological well-being among American adolescents after 2012 and links to screen time during the rise of smartphone technology. *Emotion* 2018; 18: 765–80.

14 Daugherty, D.A., Runyan, J.D., Steenbergh, T.A., Fratzke, B.J., Fry, B.N. and Westra, E. Smartphone delivery of a hope intervention: Another way to flourish. *PLoS One* 2018; 13: e0197930. https://doi.org/10.1371/journal.pone.0197930, accessed 25 August 2020.

15 Harris, D. *10% Happier: How I Tamed the Voice in My Head, Reduced Stress Without Losing My Edge, and Found Self-Help That Actually Works – A True Story.* New York: It Books, 2014.

16 Kelly, B.D. Mind your mental health during this most psychologically difficult phase of pandemic. *Irish Independent*, 25 August 2020.

17 Kelly, B.D. *Coping with Coronavirus: How to Stay Calm and Protect Your Mental Health: A Psychological Toolkit.* Dublin: Merrion Press, 2020.

18 Pinker, S. *Enlightenment Now: The Case for Reason, Science, Humanism and Progress.* London: Allen Lane, 2018. See also: Rosling, H., Rosling, O. and Rosling Rönnlund, A. *Factfulness: Ten Reasons We're Wrong about the World – And Why Things Are Better Than You Think.* London: Sceptre, 2018.
19 Bregman, R. *Humankind: A Hopeful History.* London: Bloomsbury Publishing, 2020.
20 Hobsbawm, J. We will still be social. *Monocle*, June 2020.

**Ten: Lose Yourself**

1 I have always loved Ozu's film *Tokyo Story* (1953).
2 Juliff, L. *How Not to Travel the World: Adventures of a Disaster-Prone Backpacker.* Chichester: Summersdale Publishers, 2015.
3 www.neverendingfootsteps.com.
4 For a taste of India, I can recommend these two books: Rajesh, M. *Around India in 80 Trains.* London and Boston: Nicholas Brealey Publishing, 2012; Prasad, A. *In the Bonesetter's Waiting Room: Travels Through Indian Medicine.* London: Profile Books/Wellcome Collection, 2016.
5 Bowlby, J. *Attachment and Loss* (three volumes). New York: Basic Books, 1969–80; Bowlby, J. *A Secure Base: Clinical Applications of Attachment Theory.* London: Routledge, 1988.
6 Csikszentmihalyi, M. (1990). *Flow: The Psychology of Optimal Experience.* New York: Harper and Row, 1990.
7 Gaffney, M. *Flourishing: How to achieve a deeper sense of well-being,meaning and purpose – even when facing adversity.* Dublin: Penguin Ireland, 2011.
8 Seldon, A. *Beyond Happiness: How to Find Lasting Meaning and Joy in All That You Have.* London: Yellow Kite, 2015.
9 Kelly, B.D. *The Doctor Who Sat For a Year.* Dublin: Gill, 2019.

**Conclusion**

1 De Botton, A. *The Architecture of Happiness.* London: Hamish Hamilton, 2006; Gilbert, D. *Stumbling on Happiness.* New York : A.A. Knopf, 2006; Gawdat, M. *Solve for Happy: Engineer Your Path to Joy.* New York: North Star Way, 2017; Power, M. *Help Me! One Woman's Quest to Find Out If Self-Help Really Can Change Her Life.* London: Picador, 2018.
2 https://www.parkrun.com/
3 Smith, E.E. *The Power of Meaning: The True Route to Happiness.* London: Rider/Ebury Publishing, 2017.
4 Ehrenreich, B. *Smile or Die: How Positive Thinking Fooled America and the World.* London: Granta Books, 2009.

# Index

**G**



**H**

**T**



**U**



**V**



**W**

CUNNINGHAM, ISABEL S
FRANK N. MEYER : PLANT HUNTER
IN ASIA .
40-752235
3 3012 00031 7032
WITHDRAWN
0813811481 2 003
M

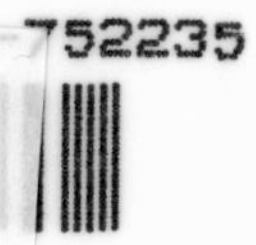

CUNNINGHAM, ISABEL S
FRANK N. MEYER : PLANT HUNTER
IN ASIA .
40-752235
3 3012 00031 7032
WITHDRAWN